GEOGRAPHY OF SIBERIA

GEOGRAPHY OF SIBERIA

2006 EDITION

Compiled and edited by

ALEXANDER N. ANTIPOV
Director
V.B. Sochava Institute of Geography
Siberian Branch of the Russian Academy of Sciences
Irkutsk, Russia

Translated by
V.G. Mikhalkovsky

RIP RESEARCH INDIA PUBLICATIONS
D1/71, Top Floor, Rohini Sector-16, Delhi-110085, INDIA

Published by
Research India Publications
Head Office: D1/71, Top Floor,
Rohini Sector-16,
Delhi-110085, INDIA
Fax No.: +91-11-27296475
Email: ripublication@vsnl.net
Website: www.ripublication.com

ISBN 978-81-904362-0-5

Price: Within India : Rs. 2200.00
Outside India : US$ 75.00

Typeset by RIP INFORMATION SERVICES
D1/71, Top Floor, Rohini Sector-16, Delhi-110085 INDIA

Printed in India at ABC Press, Naraina Indl. Area Ph-II, New Delhi-110028

Preface

This scientific monograph presents the results from geographical studies of Siberia. Its sections outline the uniqueness of this giant region characterized by ultracontinental climatic conditions, high biodiversity, the preservation of cultures of the indigenous peoples, and by a world level of natural resource potential.

Over the past five decades, the scientists of the V.B. Sochava Institute of Geography SB RAS have been engaged with the study of the natural phenomenon. Their research builds on a state-of-science theoretical foundation, the doctrine of geosystems, and on high-technology investigation methods. In the volume, emphasis is given to the advancement of cartographic representation using GIS technologies and remotely sensed data.

The reader will be introduced to the specific characteristics of Siberia's biota, soil and landscape diversity, and to the history of population formation. Particular attention is given to assessment of the region's natural resources, as Siberia has long been natural treasury of Russia and has an international significance at present.

The book is intended primarily for the international reader, specialist geographers, economists, policy-makers, as well as for all those, who are seeking for a new knowledge of the world surrounding us.

December 2006

Editors

Doctor of Geographical Science A.N. Antipov
Doctor of Geographical Science L.M. Korytny
Doctor of Geographical Science V.M. Plyusnin
Candidate of Geographical Science A.I. Shekhovtsov

Contents

From The Editor and Compiler

Siberia is a unique natural entity, having no equal worldwide as regards its size, ultra-continental climatic conditions, extremely low population density, including aboriginal ethnoses, as well as its immense natural resources. Such a combination of natural, social and economic conditions lends an unprecedented aspect to the scenery of this region that might be still scantily known to the international reader. On the other hand, the geographical space does inherently not obey any administrative or state borders or political structure alone. First and foremost it is a complex entity that lives by definite laws inherent only in it, and an understanding of these laws is the cause of scientific geographers. And our book is guided just by such an ideology.

This collection of scientific papers presents the latest research results on Siberia obtained by geographers of the V.B. Sochava Institute of Geography at the Siberian Branch of the Russian Academy of Sciences that was established in 1957 at Irkutsk. The main themes discussed in this book are intended to introduce the international reader to the distinctive characteristics of the natural environment, population and economy of Siberia, organization and development patterns of the geographical environment, policies and practices of natural resource management, and their protection and reproduction for the Siberian conditions that still remain poorly explored and developed by Man.

The book outlines a new line of scientific geographic enquiry, the theory of geosystems, that was pioneered and elaborated at the Institute by Academician V.B. Sochava (1905-1978) and has gained general recognition in Russia, and in other countries. The central element of the doctrine of geosystems is the study of the natural environment dynamics which opens up a new avenue of attack on the problem of understanding the workings of human influence on the structure and functioning of geosystems and helps unravel the mechanism of man-made impacts on Nature.

Of considerable interest to the international reader will be the integrated mapping of the natural environment, economy and population of different regions in Asian Russia and neighboring countries. The Ecological Atlas of the Irkutsk Region includes 166 unique cartographic products. At the heart of these efforts are the theory, principles and methods of cartographic modeling, a high information capacity of the maps, and real-time computer-aided mapping using aerospace information. Various research programs and projects, primarily in the Baikal region, are provided with cartographic support.

The Siberian school of geographers is concerned with issues related to population geography. The book presents a number of original ideas of the formation of population in the country's eastern regions. It is shown that the creation of an original nucleus of old inhabitants is of crucial importance for the entire process of population formation in general, and the role of Siberia's aboriginal peoples in the creation of modern population is revealed.

An impressive component of research is related to the issues of economic geography and nature management geography in the country's eastern regions. They include general issues of society-nature interaction, manufacturing locations, pioneer development of new regions, agrogeography, georesource management, economic regionalization and the study of lower-level administrative districts, and research in political geography.

The book outlines the results of Russian-German cooperation on the creation of a unique (for the Russian Federation) landscape planning methodology as a basis of sustainable development of the territory. Significant progress has been made in a number of branches of geographical science, such as landscape geochemistry, soil geography, biogeography, geomorphology, hydrology, glaciology, and climatology.

The book offers information about the mammoth reserves of natural resources in Siberia and unravels the key problems of tapping them, such as the accessibility for transport facilities, and high vulnerability of the natural environment.

This volume would permit the international reader to see with the professional geographers' eyes the novel region, experience Siberia's boundless expanses, and better understand the people living here. It is the hope of the authors that the book would throw one further little bridge thus opening up new vistas for fruitful cooperation between countries and peoples.

Director of the V.B. Sochava Institute
of Geography SB RAS, Professor
ALEXANDER N. ANTIPOV

PART I

INTRODUCTION

CHAPTER

1

Siberia and its Geographical Research

A.N. ANTIPOV, L.M. KORYTNY AND V.M. PLYUSNIN

In this book we will concern ourselves with the natural phenomenon, Siberia, a region that has not been adequately addressed in the international literature. We outline the peoples living here, with special emphasis on natural resources. Considerable attention is given to the men of science, who dedicated their lives to the study of Siberia.

NATURAL CONDITIONS

Siberia occupies most of Russia, stretching from the Ural Mountains (60° E) to the mountains skirting the Kolyma river basin (166° E), and from the shore of the Arctic Ocean (77° N) to the southern Altai ranges (48° N). Its extent from north to south is as long as 3500 km, and from west to east more than 7000 km [1]. The total area of Siberia is over 10 mil. sq. km (Fig. 1, 2).

Climate. Siberia lies in moderate to high latitudes of northeastern Eurasia, within the moderate and cold climate zones. Such a geographical location determines its key natural conditions. It is separated by many thousand kilometers from the Atlantic Ocean, and the high mountain ranges, skirting Siberia on the southern and eastern side, act to prevent the warm, wet air masses from penetrating from the seas in the Indian and Pacific Oceans. Dry, cold Arctic air masses are able to penetrate fro the Arctic Ocean deep into Siberia. For that reason, most of Siberia has a sharply continental climate. A most characteristic feature of Siberia's climate is an impressive temperature drop between a warm and a cold season. The difference between average temperatures of the coldest and warmest months varies from 35° to 65° and even reaches 105° in the Yakut north [2]. Siberia's climate continentality is also manifest by significant air temperature amplitudes over a 24-hour period, between day and night.

Figure 1: Siberia on the globe.

During a cold season, Siberia's territory is dominated by the Asian High. Its onset corresponds to October, and it occupies the whole of East Siberia by the end of autumn. The highest atmospheric pressure is observed in the Transbaikalia in winter, 1025-1035 mb. As soon as the High occurs, wet, warmer air masses cease to penetrate here from neighboring territories. Therefore, Siberia in the winter time is dominated by a clear, nearly cloudless, yet very cold and dry, weather.

Figure 2: The physiographical map.

In April, atmospheric pressure over Siberia starts to decrease, so that the Asian High gradually disappears. This is because the landmasses get warm faster than the oceanic water surface. By mid-summer, air pressure nearly throughout North Asia is above normal (1000-1010 mb) – wet air masses are arriving in Siberia from the Atlantic. Because of an abundance of solar irradiation (due to the small amounts of clouds), the summers in Siberia are warmer than along the same latitudes of the European part of Russia. Sometimes, however, relatively cold winds blowing from the Artic cause summer temperatures to decrease, with light frosts occurring in June and August.

In Siberia, yearly mean air temperatures are almost everywhere below 0° C. The harsh Siberian climate is characterized by very low winter temperatures (from -35 to -50° C) and by the long, cold winter. Furthermore, the winter in east Siberia is by about 20° colder when compared with West Siberia. Siberia over the last two or three decades has been experiencing a rise in yearly mean air temperature (0.03-0.08 °C/year) due to the abatement of the winter colds. By calculating the climatic fluctuation cycles, it was possible to predict the possible air temperature changes [3]. The currently observed warming may well continue into the year 2020, but the time interval 2040-2060 will see a cooling implying a decrease in mean temperatures by 0.5° C against the 1950s-1970s.

Most of atmospheric precipitation that occurs over Siberia in the form of rain and snow, is transported by air masses from the north or north-west. The amount of precipitation decreases toward the east. The distribution of atmospheric precipitation is uneven in different seasons. Up to 80% of the annual amount of precipitation correspond to a warm season; this is especially true of July and August. In the mountains, the amount of precipitation exceeds 1000 mm per year, in the taiga zone of West Siberia the amount is 400-600 mm, in the taiga of East Siberia 300-400 mm, in the forest-steppe of southern Siberia as much as 300 mm, and in the tundra zone as much as 200 mm.

The wide occurrence of permafrost in Siberia is also explained by continental climatic characteristics (Fig. 3). It is the result of a long-lasting, deep cooling of the surface of the Earth's crust. Permafrost was produced as early as the glacier period when the harsh, snow-deficient, sharply continental climate was still more pronounced than at the present time. The formation of permafrost was caused by losses of large amounts of heat in anticyclonic conditions of a cold period, and by a deep freezing of earth materials, and their total unfreezing is impossible over the summer time. In this manner, a gradual "accumulation of the cold" had been proceeding over the course of hundreds and thousands of years, resulting in a decrease in permafrost temperature and an increase in the permafrost thickness. Continuous permafrost is typical of the Siberian north, with the permafrost thickness reaching 300-600 meters and the temperature at a depth of 10 meters ranging from -10 to -12° C. Permafrost including islands of taliks can be found further to the south. The permafrost thickness in this case is as small as 25-50 meters, and the permafrost temperature rises to -1 or -2° C [4]. Only permafrost islands can be found in the southern part of Siberia. They are represented by small patches of permafrost in topographic depressions, and on northward mountain slopes. The permafrost thickness in the south is 5-10 meters. As the permafrost is an acquiclude, it acts to enhance the surface runoff, makes deep-seated erosion difficult and contributes to lateral erosion thus favoring the development of cryogenic landforms.

Relief. Siberia is a country with quite a varied relief. Its western part is occupied by the extensive West-Siberian Plain, while the Middle-Siberian upland (one of the most significant uplands on the globe) occupies Middle Siberia. A broad zone of high mountains (Altai, Sayan, Pre-Baikalia's and Transbaikalia's mountains, and the mountains in the northeastern part of Russia) runs along the southern and eastern boundaries of Siberia.

The West-Siberian Plain is one of the largest accumulative low-land plains on the globe. Differentiated descents of the West-Siberian plate during the Mesozoic period had led to the fact that it is now dominated by the accumulation processes of friable sediments. The plain is shaped as a step-like circle of northward exposure, toward the coast of the Kara Sea. Seaside accumulative plains of maximum transgression lie in the northern part of the West-Siberian Plain and occupy the area along the left-bank of the Yenisei, most of Gydansky Peninsula and Yamal peninsula, and the Taza, Pura and Nadyma river basins.

The glacier and water-glacier plains lying further to the south are characterized by the most dissected gently rolling topography, and they spread in the form of a broad zone from the Ural to the Yenisei range.

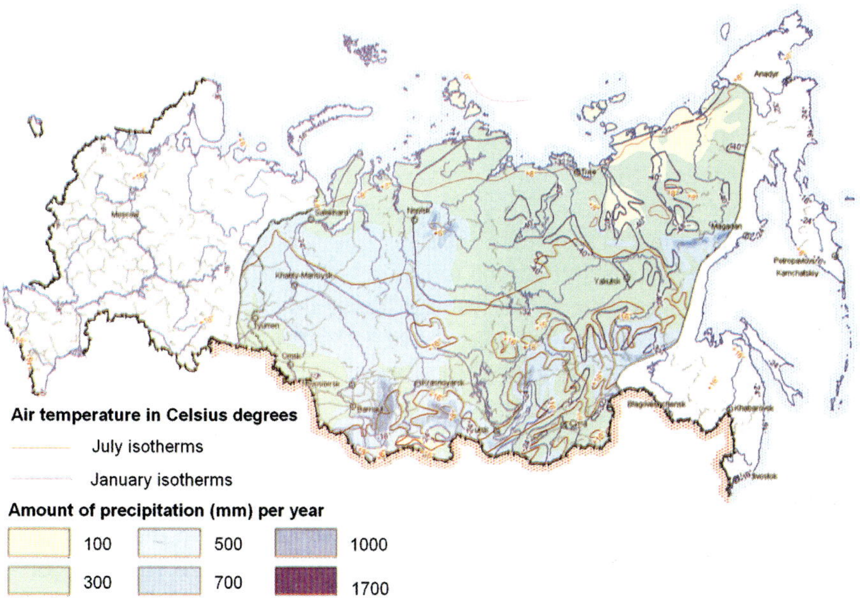

Figure 3: Map of mean air temperatures.

The western part of the Siberian platform is occupied by the Middle-Siberian upland. It is characterized by significant elevation (as high as 1700 m), and by a combination of gently undulating step-like topography of the interfluves, with deeply incised steeply sloping river valleys. The long period of continental development of the territory is responsible for a predominance of denudation topography, while its step-like appearance was caused by the recent uplifts and alternation of surface-forming rock and earth materials differing in its resistance to weathering. The large number of terraces (as many as nine) along the river valleys provide evidence of the remote origins and repeated structural highs (tectonic uplifts) of the territory.

The mountain regions of Siberia's southern and eastern provinces are distinguished for their complicated geological structure and by extremely varied landforms. The alpine high-mountain relief occurs in the elevated Altai and Sayan ranges, in the Stanovoi upland, Suntar-Khayat, the Chersky range, and in others at higher than 2000-2200 m elevation. Such tracts show a significant depth of dissection, large elevation differences, a predominance of steep hillsides, and narrow watershed ridges with peaks difficult of access. A critical role in hillside denudation was played by the old mountain glaciers that produced numerous circles, troughs, and moraines. The highest peaks are: Belukha Mountain on Altai – 4506 m, Munku-Sardyk Mountain – 3491 m in the Sayan Mountains, Kodar Range – 3073 m, and Muss-Khaya Mountain in Suntar-Khayata – 2969 m. Among the modern exogenous processes are mountain creep, taluses, nivation, glacier exaration, avalanches, debris flows, and river erosion.

Glaciation. Siberia's ancient glaciation in the Quaternary period had been recurring in the past. Noteworthy are the Samarovo (maximal, Tazovo, Zyryanovo and Sartan (mountain-valley) glaciations. The glaciation cores in Siberia were provided by the Ural-Novaya Zemlya, the Putoran Plateau, Taimyr and the Anabar mass. Mountain-valley glaciation occurred in the southern Siberian mountains and in the North-East.

Siberia's modern glaciers can be found in high mountains; they are small in their area and mostly occur in circles and along the valleys. For instance, the Western and eastern Sayan Mountains have 159 glaciers, with an area totaling 34.1 sq. km. The Kuznetsk Alatau includes about 90 glaciers with a total area of 7 sq.km. In the Stanovoy Upland (the Kodar range) there are 39 glaciers with an area over 15 sq. km.

Waters. The territory of Siberia is rich in surface and subsurface water. Almost all rivers of Siberia refer to the watershed of the Arctic Ocean, and most of them are fed by snowmelt water in the spring, and rainfall in the summer and autumn seasons. Especially large is the role of rainfall in the summer for the rivers of Siberia's mountainous part, with a significantly smaller amount of atmospheric precipitation than on the plains. The amount of runoff for a warm season constitutes 80-90% of the annual amount. Toward the end of the autumn when first severe colds set in, all rivers of Siberia freeze up. Small rivers of the northern regions are the first to free up; the rivers of the tundra and taiga zones freeze up in October, while the ice cover appears on the southern rivers of Siberia in early November. The largest Siberian rivers, Ob, Yenisei, and Lena, that transport warm waters from the south, freeze up somewhat later than their tributaries. A large number of small rivers freeze up to the bottom by the end of the winter. In East Siberia, on many rivers there occur icing phenomena which are mainly caused by subsurface water emerging on the surface or by the freeze-up of the river water expelled by the anchor ice. Break-up of the Siberian rivers starts in April in the south; the ice disappears during May on all rivers of the taiga zone, and only in the north it continues till June.

The Ob, Yenisei and Lena are the largest rivers of Siberia. The Ob is one of the world's largest rivers. The area of its basin is about 3 million square kilometers, and its length from the outflow of the Irtysh is 5410 km. The annual flow rate of the Ob is about 400 cu km. In Russia, the Ob ranks 3rd from the top in water abundance, next to the Yenisei and Lena. The river flows into the Ob Guba, a typical estuary. The Ob and Irtysh with their tributaries within West Siberia are typical plains rivers with small gradients and calm flow.

The area of the Yenisei river basin is slightly more than 2.5 mil sq. km. The river has its origin in the mountains of Tyva and flows into the Yenisei Bay of the Kara Sea. The Upper Yenisei is an impetuous mountain river with large gradient. In its middle part, where the river flows just along spur of the Middle-Siberian Upland, and its channel is home to many a large rapids, and its flow has high speed. Only the Lower Yenisei flows quiet. The river if 4092 km long, and its annual flow rate is about 624 cu km [5]. It is the most water-abundant river in Russia.

In Russia, the Lena occupies the third place in the area of its basin (2490 sq. km), and the second place in the flow rate, next to the Yenisei. The length of the Lena reaches 4400 km, and its annual flow rate is 536 cu km. The Lena has its source on the western slope of the Baikalsky Range, and its upper part represents a typical mountain river. The Lena, downstream of its outflow into the Laptev Sea, acquires the character of a large plains river. In the place of its inflow into the Laptev sea, it forms Russia's largest delta with an area of more than 32 thou sq. km.

Siberia's largest lakes are Lake Baikal (31500 sq.km the area of its water surface), Taimyr (4560 sq. km), and Chany (2000 sq. km). The West-Siberian Plain includes about one million lake with an area totaling more than 100 thou sq. km.Lake coverage varies from 1-1.5 in the south to 2-3% in the north. The large number of lakes is caused by the existence of flat surfaces, and by weak drainage of the territory. Lakes occur in the watershed areas as well as in the river valleys. The North-Siberian and Central-Yakut lowlands also have a high lake percentage, with small, shallow thermokarst lakes predominating. Large lakes in the depressions of a glacier-tectonic origin can be found on the Putoran plateau: the Khantaiskoye, Kheta, Lama and Pyasino lakes. They occupy narrow depressions with steep slopes, resembling Scandinavian fjords.

Lake Baikal is one of the world/s largest lakes, the volume of its water is 23 thou cu km (occupying the first place in the world); the lake is 664.6 km long, and its mean width is 49.6 km (max = 86.2 km, min = 28 km), its maximum depth is 1637 m, and its mean depth is 730 m (Fig. 4). The length of Baikal's shoreline is 2 100 km. Baikal's water is characterized by low content of mineral salts (96.4 mg/l), water transparency is 40 m, and the lake is 20-25 million years old. The lake receives 518 permanent streams (the largest is the Selenga 1024 km long, its annual flow rate is 16.7 cu km, and its basin is 447 thou sq. km), and the Angara is the only river flowing out from the lake. The total yearly water input to Baikal is 71.16 cu km (9.29 cu km atmospheric precipitation, 58.75 cu km correspond to the rivers, 2.3 cu km to subsurface water, and 0.82 cu km comes from condensation on the lake's surface). The area of the entire watershed basin of Lake Baikal is 571 thou sq km. Baikal has a unique flora and fauna. Numerous relict animal and plant species inhabit the lake. For that reason, Lake Baikal was declared as a World Natural Heritage site in the year 1996.

A giant reservoir of water is provided by the peatlands of West Siberia. The mean swampiness of the plain makes up abou 30%; in the forest-peatland zone it is about 50%, and in separate areas (Surgut Polesie, Vasyuganie, and the Kondinskaya lowland) it reaches 70-80%. The world's largest swamp is the Vasyugansk bog with an area of 53 thou sq. km. The total area of peatland in Siberia is estimated at 1 million square kilometers, with the reserves of peat amoung to about 90 billion tons in an air-dry state; moreover, the peat occurring in the peatlands of West Siberia contains no less than 1000 cu km of water.

Figure 4: Lake Baikal.

Soils. The subdued plains relief of West Siberia is favorable to a clearly pronounced zonality, andto a high degree of hydromorphism. The zonal types of soils are tundra-gleyey, podzolic (under coniferous forest), soddy-podzolic (under birch forests), eluviated and degraded black earths or dark-gray podzolized soils (under isolated birch stands), black earths (occurring in thin tongue-shaped layers), and dark-chestnut soils occur in relatively well drained territories. In West Siberia, semi-hydromorphic soils occupy large areas not only in the tundra and forest-tundra but also in the forest-boggy and forest-steppe zones.

Soils in east Siberia occur predominantly on bedrock eluvium; therefore, they normally include granitic soils of a small thickness. Soil formation is proceeding in conditions of shallow occurrence of permafrost. In the far North there occur Arctic-tundra, gley tundra and sub-brown soils. Frozen taiga soils are evolving in the northern half of the taiga zone. Their chief features are associated with subsurface layers of permafrost that create conditions for the impermeable water regime thus making the exit of salts difficult. In the south, where the permafrost occupies small areas, soddy-podzolic soils prevail, while the forest steppe is dominated by gray forest soils and chernozems occurring in thin layers.

Vegetation. The distribution of vegetation cover in West Siberia obeys the zonality law (Fig. 5); in East Siberia, zonality is traceable less as clearly. This is due to significant elevation of the territory resulting in high differentiation of the natural conditions, which makes the manifestation of zonality difficult.

Figure 5: Map of Vegetation.

The tundra zone occupies the northernmost part of Siberia, adjacent to the seas of the Arctic Ocean. Tundras are stretched in the form of a continuous strip from the Ural Mountains to the Kolyma river basin 500-600 km wide. Characteristic features of the tundra vegetation is a wide occurrence of mosses and lichens, an abundance of small shrubs, and no forests at all. Further to the south, the vegetation cover is becoming continuous, and number of bare (devoid of vegetation) patches decreases, and the Arctic tundra gradually gives way to the moss-lichen tundra. Still further to the south, the number of shrubs gradually increases; they become taller (as high as 1.5 meter) to form continuous thickets not only along the valleys protected against wind, but also in the interfluve areas. Shrubs include wild rosemary, bog whortleberry, and red whortleberry. The surface is clad with a mantle of green and sphagnum mosses. There have already appeared a rather large number of herbaceous plants, such as fescue, meadow-grass, cinquefoil, and forget-me-not.

Further to the south of the woodless tundra in Siberia is a narrow strip of forest-tundra in which tracts of tundra vegetation alternate with sparse forests and raised peat bogs. Among timber species, larch is predominant (Siberian larch to the west of the Pyasina river, and Daurian larch further to the east). West Siberia is dominated by sparse spruce-larch forest stands. In the southern part there occur Siberian spruce and white birch. Pinus Pumila (Pall.) Regel occurs in the easternmost part.

The taiga zone in Siberia stretches over an area more than 7 million square kilometers. Not a single country in the world has such an immense area occupied by forests. Siberian forests are dominated by coniferous tree species (Fig. 6), which is explained by the present climatic characteristics of Siberia's forest zone as well as by the peculiar history of formation of its landscapes over the course of the Quaternary period. A cooling that was followed by Pleistocene glaciations, caused broad-leaved forests to disappear, which had had a wide occurrence throughout Siberia. In the western part of Siberia, broad-lived forests give way to dark-coniferous forests consisting of spruce, fir, and Siberian stone pine. In East Siberia, with its drier climate and the presence of permafrost, larch forests appeared. In the taiga forest, the bulk of dark-coniferous species includes four spruce species, three fir species, the Siberian stone pine, and tree-like shrubs, i.e. Pinus Pumila (Pall.) Regel. Light-coniferous species include common pine, and four larch species [6].

The taiga zone of Siberia is categorized into three subzones according to the character of its vegetation: northern, middle, and southern.

The northern taiga zone in West Siberia occupies flat plains of the Nadym, Pur, Taz, Turukhan and Lower Ob river basins. In the interfluves there everywhere occurs permafrost, with woodless hillocky peatland predominating. Sparse larch lichen forests occur long the river banks and on elevated ridges, with the inclusion of birch, as well as pine stands and swamped larch-spruce-Siberian stone pine forests. The share of forests in the total area of the subzone averages slightly more than 20%.

The middle taiga zone is particularly typical of West Siberia. Its width reaches 500-550 km, and forests occupy about one half of its area. Special mention should be made of the most characteristic pine stands on white moss and on podzolic soils along dry ridges (ouvals) and riverside crests, and pine stands on green moss at wet locations. Riverside crests are clad with the Siberian stone pine – spruce green moss taiga. It also includes fir, pine, and birch. In addition to shrubs of willows, common bindweed and alder, underbrush also includes bird cherry, mountain ash, elder, juniper, and honeysuckle. Forests of the middle taiga subzone in West Siberia are often modified as a result of forest fires and timber felling; in some places, secondary birch stands account for about 25% of the forested area.

Figure 6: The Taiga.

In East Siberia, the middle taiga subzone is represented by forests including larch (Siberian – in the west, and Daurian in the east). There also occur secondary birch and birch-aspen tree forests occupying as much as 18% of the area. Growth of wild rosemary, red whortleberry, and bog whortleberry are most common here under the forest canopy. Only in the areas along the Yenisei do dark-coniferous forests of fir, spruce and Siberian stone pine occur, but they occupy a relatively small area. Swampiness here is significantly less than in West Siberia, and bogs occur mainly on flat, weakly drained interfluves.

The southern taiga subzone of West Siberia is dominated by coniferous and birch-tree forests. It is characterized by a smaller number of bogs, with dark-coniferous forests (especially in its eastern part) of fir, spruce and Siberian stone pine predominating. That Siberian fir prevails in dark-coniferous forests is a typical attribute of the southern taiga. Birch-tree forests are also widely distributed – they occupy as much as 40% of the forested area.

The southern taiga subzone of East Siberia includes forests of the Angara and Upper Lena river basins. Significant areas here are occupied by pine stands, shrubs or grassland. The Angara region is home to a particularly large number of pine and pine-larch forests – this is due to moderate atmospheric humidity, and to the wide occurrence of sandy terraces. Further to the east, in the Lena and Aldan river basins, they give way to the Daurian larch taiga. Dark-coniferous fir, spruce or Siberian stone pine forests can be found in the most elevated, humidified interfluve areas, and along some river valleys.

The zone of small-leaved forests stretches in the form of a narrow lane (as narrow as 50 to 200 km) along the southern edge of the taiga zone. Its vegetation cover includes mainly aspen-birch tall forests on grass-clad soils. About 55% of the forested area correspond to birch stands with wood sorrel as well as birch stands with wood sorrel and horsetail, on the grass mantle composed by Aconitum, goutweed, Calamagrostis epigeios (L.) Roth., and other meadow plants. In addition to birch stands, the zone includes pine groves, as well as small stands of Siberian fir, Siberian stone pine, and aspen-tree forests on grass-clad soils.

The forest-steppe zone in West Siberia stretches in the form of a narrow strip (150-300 km) from the Ural Mountains to Altai, and is divided into two subzones: northern, and southern. In the northern forest-steppe there are a large number of bogs, swampy meadows, isolated birch stands, and coppices. Usually, isolated stands occur at sinks with formation of solods, heavily leached, degraded solonetz or dark-gray forest soils. They consist of white and common birch with the inclusion of aspen-trees and dwarf willows in undergrowth stands. Trees are as high as 10-12 meters. Open woodless spaces were previously dominated by steppe meadows, and they are currently ploughed up almost everywhere. Floodplains of the river valleys are covered with large-grass meadows, with meadow-sweet, bent, sedge, and clover predominating.

In the subzone of the southern forest-steppe, isolated tree stands occupies a still smaller area (as small as 5% of the territory). Open spaces were previously dominated by meadow steppes. Their composition mostly included European feather grass and fescue (xerophilous herbaceous grasses on soddy soils). Today, they are also ploughed up. A characteristic property of the southern forest-steppe is the presence of ribbon-like pine groves) consisting of excellent timber) as long as 200 km and as wide as 25 km. Their distribution is caused by the sandy soil medium, and by the shallow occurrence of subsurface water. Such pine groves include large numbers of typically forest species: mountain cranberry, bilberry, Pyrola, Linnaea borealis L., cranberry, as well as lots of true "steppe dwellers": fescue, wormwood, Caragana jubata, and kiptsa.

The steppe zone in West Siberia is currently almost totally ploughed up. Natural vegetation has persisted only in small virgin-soil areas (Fig. 7). Two subzones may be identified in this case: feather-grass – forbes and feather-grass

– fescue steppes. In topographic depressions, some places show the presence of thickets of low steppe shrubs: spirea, Tartar honeysuckle, Caragana jubata, cherry, and Labornum. In places with saline groundwater occurring near the surface, shrubs give way to wormwood and salt-loving plants: teresken, and sea lavender.

Figure 7: Steppes.

In east Siberia, relatively small patches of forest-steppes and steppes can still be found mainly in intermountain depressions, and on piedmont plains.

Fauna. The fauna of the tundra zone is much poorer and more homogeneous when compared with the fauna of Siberia's other zones. The most typical inhabitants of the Siberian tundra are small mice-like rodents: lemmings, and voles. The polar owl and tundra partridge live in the tundra throughout the year. White polar bear can be encountered along the seaside of the Arctic tundra. Typically, in the summer time the tundra is inhabited by the following birds: geese, ducks, gulls, sandpipers, loons, and birds of prey hunting them: the fluffy-legged buzzard, peregrine falcon, and bald eagle. Permanent predators in the tundra are: Arctic fox – it is highly values for its snow-white fur-skin and is therefore the most important hunted animal of the tundra zone. In addition to the Arctic fox, the tundra is also home to other beasts: the wolf, fox, ermine, weasel, and otter. The largest animal of Siberia's tundra zone is reindeer. Their numbers are few to date, and they constitute an important sector of the economy pursued by the local population of the tundra in the form of reindeer breeding. The numerous rivers and lakes of the tundra zone abound in fishes. One can find such valuable species as whitefish (powan), muksun, salmon-trout, Siberian sturgeon, starlet, nelma, Siberian cisco, loach, and smelt.

The commonest animals of the forest tundra zone are: the ermine, lemmings, voles, sgrew, fox, white hare, glutton, and brown bear. In the winter time, visitors arrive from the tundra: the reindeer, wolf, and Arctic fox. Birds in great numbers arrive from the south in the summer: the sandpiper, duck, gray crane, thrush, cuckoo, peregrine falcon, horned skylark (ryum), pine grosbeak (Pinicola enucleator), and others.

As regards large animals, the commonest for the taiga are the brown bear, glutton, lynx, moose, Manchurian deer, musk deer, and bighorn sheep (Fig. 8, 9).

Smaller (yet typical of the taiga) beasts in Siberia include fox, sable, least weasel, kolinsky, polecat, marten, otter, and badger (in the south). The taiga particularly abounds in various rodents. The squirrel is of the greatest hunting value. Apart from the squirrel, one can encounter everywhere chipmunks, white hares, flying squirrels, and various mice-like rodents. Most of the birds are dendrophilous birds. Their typical representatives are the Tetraonidae: the cock of the wood, and the hazel grouse, as well as the red-headed jay, Siberian jay (Perisoreus infaustus (Linnaeus), nutcracker, black woodpecker, three-fingered woodpecker, crossbill, pine grodbeak, wax-wing, and nuthatch. The Passeridae can be found in especiallu great numbers: the thrush, tit, starling, bullfinch, chaffinch, bramble finch, and oriole. In the summer, the region is abounding in water fowl: geese and ducks, while meadows and bogs are host to sandpipers, pewits, curlews, comcrakes, and yellow wagtails. The taiga provides a habitat of many species of insects connected with coniferous forests. A variety of silkworm species (Siberian and pine silkworm, and gypsy moth). Nun-moth, pine hawk-moth, pine cutworm, pine moth, pine sawfly, large wood-wasp, stag beetle, and bark beetles.

The unique environmental conditions of Lake Baikal are responsible for the high endemism of its flora and fauna.

Lake Baikal is known for its 1550 animal species, and 1085 species of vegetable organisms, about 1000 species of which are endemic (83%). Baikal's shore area if home to more than 2 thousand vascular plants, the overwhelming majority of which are categorized as endemic and relict. Of the 56 species of fishes living in Baikal, 27 species are endemic, and 17 species are commercially caught. There are 236 bird species in the Baikal area, including migrating birds, as well as 30 species of game animals. The biomass of fatfishes (golomyanka) on Baikal is 150 thousand tons, and the biomass of the other fishes is 74 thou t (with 25-39 t corresponding to omul (Baikal cisco) (Fig. 10); about 2 thou t are caught every year). About 75 thousand head of seal live on Baikal.

The animal world of the forest-steppe has a transition character. Various rodents are most typical of the forest-steppe, such as the ground squirrel (souslik), hamster, jerboa, voles, and steppe lemming. Roe and white hare, and also predators: wolf, corsak fox, polecat, ermine, least weasel, badger, and various fox species live here. In connection with the abundance of rodents, the Siberian forest-steppe area is home to a great number of birds of prey: the steppe eagle, tailbone, kestrel, and harrier. Furthermore, common to this area are the many-colored woodpecker, skylarks, gray partridge, black grouse, tailbone, turtle-dove, magpie, and oatmeals. As regards the herpetofauna, mention may be made of the common triton, and sharp-nosed frog.

Figure 8: Bear.

Figure 9: Sable.

Most steppe animals feed on steppe grasses and forbes. This area abound sin various smaller-size rodents: earth hare, feather-legged jerboa, red-cheeked ground squirrel, gray and herding voles, steppe tree creeper, mole vole, and steppe lemming. They are hunted by steppe predators: light steppe polecat, corsak fox, common fox, wolf, least weasel, ermine, and badger. Lots of rodents are hunted by birds of prey: the steppe eagle, buzzard, lesser kestrel, Montagu's (meadow) harrier, and short-eared owl. The steppe zone is also characterized by the presence of the white-winged skylark, little bustard, yellow-hammer, and other birds. Insect-eaters are represented by the long-eared hedgehog. The herpetofauna also includes Agkistrodon halys pallas, grass snake, sand lizard (Lacerta agilis), and green toad.

Figure 10: Omul (Baikal Cisco).

THE POPULATION OF SIBERIA

The immense area of Siberia, constituting 56.5% of Russia's territory), its uneven population distribution, combined with the severe natural conditions, are responsible for the differing levels of economic and social development in its different parts, as well as a sharp differentiation of territorial organization of the production and settlement. The population living in Siberia (22 052 thousand, or 16.8% of the country's population, as of 1 January 2005) has to cope with challenging issues relating to the utilization of natural resources and economic development. The territory is unevenly populated, and this factor continues getting momentum. Siberia is exceeded in population density by the European part of Russia by a factor of 11 (and by a factor of 20 against the average figure worldwide) (Fig. 11).

The population is distributed among 22 subjects of the Russian Federation (Table). The centers of the economic and human potential are represented by 174 cities and towns, and by 326 urban-type settlements, with the overwhelming majority of the population living there, 71.6%. The urbanization process has been evolving at a great pace and was accompanied by the emergence of new towns and urban-type settlements. The largest cities are Novosibirsk (1406 thousand), Omsk (1143), Krasnoyarsk (917), Barnaul (631), Irkutsk (583), Novokuznetsk (563), Tyumen (538), and Kemerovo (523).

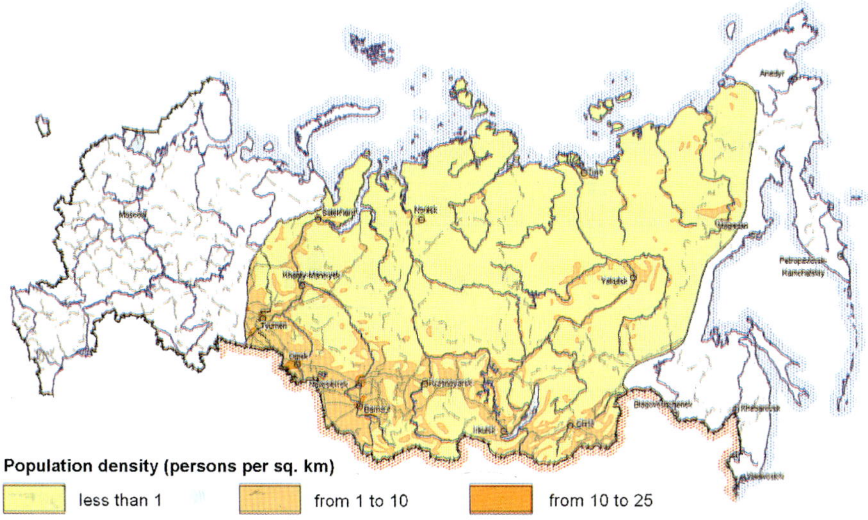

Figure 11: Population density map.

84.4% of Siberia's poly-ethnical population (over 100 nationalities (or ethnical groups) according to the 2002 census) corresponds to the Russians. As far as the other nationalities are concerned, 2.9% correspond to the Ukrainians, 1.8% to the Tatars and Siberian native peoples (with 1.5% corresponding to each of the largest of them, the Buryats (Fig. 12) and Yakuts).

Remarkably, the southeastern part of Siberia shows the Russian minimum of life expectancy, or 3-4 years below the average figure for Russia. The Siberian area of minimum life expectancy, with its center in Tyva, includes the Republic of Altai, the Republic of Buryatia, and the Irkutsk and Chita regions.

The long-lasting period of population formation in Siberia through an expanded natural reproduction and immigration has given way to a dwindling reproduction and emigration. From 1991 to 2005, the population in the region decreased by 1 432 residents (5.6%), so that the today's Siberia has the same population as two decades ago. Such negative demographic manifestations as natural decline, and gradual ageing because of the low birth rates, are the result of the dwindled reproduction of population inherent in developed societies, while the high mortality and the shortened duration of life result from a peculiar Russian quality of life. There is a gradual (from west to east) increase in the relative migration-cased loss of population. Migration losses are aggravated by unfavorable structural characteristics of emigration (young, educated groups of people).

Table

The population of Siberia

Region	Permanent population as of 1 January 2005	Change in population for 1991-2005		Mean natural population growth for 2000-2004
		thousand	%	
Siberia	24052.4	-1432.0	-5.62	-3.38
West Siberia	14677.5	-445.2	-2.94	-3.66
Republic of Altai	203.9	8.1	4.14	1.31
Altai Territory	2565.6	-89.5	-3.37	-5.11
Lemerovo region	2855.0	-245.4	-7.92	-7.53
Novosibirsk region	2662.3	-85.4	-3.11	-5.16
Omsk region	2046.6	-118.0	-5.45	-4.54
Tomsk region	1036.6	-49.7	-4.58	-3.59
Tyumen region	1315.1	-28.2	-2.10	-3.26
Khanty-Mansi Aut. Distr. – Yurga	1469.0	140.8	10.60	6.03
Yamal-Nenets Aut. Distr.	523.4	22.1	4.41	7.23
East Siberia	8424.2	-816.2	-8.83	-3.77
Republic of Buryatia	969.2	-83.9	-7.97	-1.72
Republic of Tyva	307.6	1.5	0.49	4.12
Republic of Khakasia	541.0	-34.3	-5.96	-5.06
Krasnoyarsk Territory	2868.6	-221.9	-7.18	-4.87
Taimyr (Dolgano-Nenetsky) Aut. Distr.	39.4	-15.0	-27.57	4.26
Evenki Aut. Distr.	17.4	-8.0	-31.50	2.24
Irkutsk region	2411.2	-258.8	-9.69	-4.72
Ust-Ordynsky Buryat Aut. Distr.	134.1	-3.8	-2.76	-0.29
Chita region	1062.2	-187.5	-15.00	-3.47
Aginsky Buryat Aut. Distr.	73.5	-4.5	-5.77	4.24
Republic of Sakha (Yakutia)	950.7	-170.6	-15.21	4.37

Figure 12: Ethnical representatives.

The region, as before, cannot boast a significant population, but if the unfavorable trends continue, this would destroy the demographic potential of some of the territories. It is important to define the strategy to "annihilate" the perception of Siberia as the appendage of raw materials and enhance the special geopolitical and geoeconomic role played by the eastern regions for the Russian Federation. Adjustments should be made to the new demographic realities as well as looking for political, economic and social answers to the challenges of the present time.

THE NATURAL-RESOURCE POTENTIAL

Siberia is a macro-region with a huge unprecedented natural-resource potential in Russia.

Mineral resources. Russia has a world lead in its fuel and energy resources, most of which are in Siberia. One third of the world's reserves of natural gas (heading the list on the globe) has been discovered in Russia, 92% of which correspond to Siberian gas. The country occupies the second place (12%) as regards the reserves of petroleum and gas condensate, and 70% of these reserves are in Siberia. The coal resources constitute about 30% of the world's resources, with ¾ of the coal discovered in Siberia.

The ore resources are as great. Russia occupies the 3rd place (7-8%, next to SAR and USA) in gold reserves, and nearly one half of gold is currently extracted in Siberia. Russia occupies also the 3rd place in the reserves of copper (about 10% of the world's reserves), and again Siberia heads the list as regards the reserves and extraction (64%). This country has the world's largest reserves of platinum and platinoids (40%), nickel (35.8%), iron (26%), titanium (18%), and zinc (about 15%), a significant part of which (except for iron and titanium) corresponds to Siberia: respectively, 90, 70, 17, 24, and 66%. A similar situation exists as regards lead, cobalt, tin, tungsten, molybdenum, and manganese, with most of their reserves concentrated in the Siberian expanses – respectively, 91, 60, 50, 43, 88, and 66% of the reserves in Russia. Siberia produces 81% of the country's primary aluminum, and here Russia occupies the second place in the world (13%, next to China); however, mainly imported raw materials are used.

The picture is also similar as regards non-metallic mineral resources. Russia's supremacy is well known as regards the reserves of diamonds, with 99% of them extracted in Russia nowadays. Siberia is also home to most reserves of muscovite mica (81%), and graphite (46.3%), as well as to a great number of deposits of building materials, precious and semi-precious stones, etc.

Water and hydropower resources. Siberia has immense river flow resources, the most valuable water resource, amounting to 2350 cu km. 2/3 of them are contained in three powerful river systems: Ob-Irtysh, Angara-Yenisei, and Lena. The reserves of such resources in Siberian lakes are larger by an order of magnitude; Lake Baikal alone contains 23 thou cu km of water. The water resources of the Siberian reservoirs are intensely utilized, the largest of which is the Bratsk reservoir, the world's second in the water volume. Siberia's natural resources of subsurface freshwater are estimated at 420 cu km / year, or nearly one third of Russia's resources.

Of the greatest importance for Siberia and for the country in general is the tapping of the huge hydropower potential of Siberia's major rivers. Currently 11 hydro-electric power stations (HEPS) are in operation: the Irkutsk, Bratsk and Ust-Ilimsk HEP on the Angra, the Sayan-Shushenskaya, Krasnoyarsk and Mainskaya HEPS on the Yenisei, the Novosibirsk HEPS on the Ob, the Ust-Khantaika HEPS on the Khantaika, the Kureika HEPS on the Kureika, the Vilyui HEPS on the Vilyui, and the Mamakan HEPS on the Mamakan, with the power totaling 24 million kW, and the mean annual generation of electric energy over 100 billion kW/hr (more than one half of the total energy generated by the country's HEPS). The construction of the major Boguchan HEPS, and of a number of other HEPS is nearing completion. 95% of the total hydropower is generated by the HEPS of the unique Angara-Yenisei cascade, where the energy generated by them is the world's cheapest.

Forest resources. Nearly one half of Siberia's entire territory is covered with forests consisting largely of coniferous species (83% of the total area). Such forests are commonly termed the taiga forests. It is the world's largest, relatively weakly disturbed forested area (nearly one half of all forests in the northern hemisphere) which performs, among other things, key ecological functions of producing oxygen and absorbing carbonic acid gas, as well as the function of biodiversity conservation. About 14% of the forested area have been organized into a special environmental protection group, including the most valuable forests of Siberian stone pine. Siberia's forests contain about 35 billion cubic meters of wood (such as coniferous wood of the highest quality). The forests are home to a great many game animal species, such as unique fur-bearing animals (sable, kolinsky, etc.).

An important role in Siberia's economy is also played by *land* and *recreation* resources. Fertile soils exist in steppes and forest-steppes of Southern Siberia, as well as in the floodplains of major rivers, such as the Ob in particular; they are able to yield good grain harvests as well as industrial crops and vegetables. In many mountainous areas, the hillsides are suitable for pasture, and in the northern tundras and forests, the soils are good for the development of reindeer-breeding. Recreation resources of international significance are concentrated in the mountains of Altai, Tyva, Sayan, as well as in the unique Baikal region.

THE HISTORY OF EXPLORATION AND GEOGRAPHICAL STUDIES OF SIBERIA

Before the advent of the Russian people in the 11th-12th centuries in Siberia, its population exceeded 300 thousand. It seems most likely that the Russians began penetrating the areas beyond the Ural Mountains eastwards when inhabitants of the old Russian town Novgorod organized their campaigns beyond the Polar Circle and the northern Ural Mountains to the Northern Sosva river basin (forming part of the Ob river system). The detachments from Novgorod used to meet there with taiga hunters and fishermen (Yugra, Mansi), and later with their northern neighbors (Samoyadi) (Fig. 13). They laid them under tribute, and as early as the mid-18th century Yugra was included in the Novgorod volosts (smallest administrative division).

Perhaps, as early as the 12th-13th centuries the Russian pomors (coast-dwellers, inhabitants of coasts of White Sea) used to look for fur-skins and walrus rookeries and visit the mouths of the Ob and Taz rivers. They began to trade with local inhabitants: the Khants, and Nenets. In the early 16th century, long-term activities of land surveyors from Moscow started. They embarked on compiling "drawings" of the Russian (including eastern) lands. The second half of the 16th century saw the final "conquest" of the southern part of West Siberia, resulting in its annexation to Russia. Noteworthy is the emergence of Siberian towns: Tyumen (1586), Tobolsk (1587), Berezovo (1593), Obdorsk (Salekhard at present) (1593), Tara (1594), and others. These towns became "outposts" for geographical studies of Siberia. As before, participants of military campaigns, followed by explorers, hunters and traders, used to prepare geographical descriptions (including maps) of the places they had traversed.

The early 17th century saw the completion of the exploration of the Upper Ob river basin, and the town of Tomsk (1604) was founded to become a leading center of eastern studies. Later on, the town of Kuznetsk (1627) was founded. In the lands of the indigenous people Shorts, the Russians saw the first mountains beyond the Ural: the Salair, Kuznetsk Alatau, and Abakan ranges, and somewhat later the Altai. The detachment guided by Pyotr Sobansky discovered Lake Teletskoye.

The exploration of East Siberia was underway at the same time. As in West Siberia, it began from the northern part. As early as 1607, the Russian explorers founded Novaya Mangazeya at the place where the Yenisei receives its left-hand tributary Turukhan. Through the watershed area, they also penetrated into the Middle Yenisei, where they met for the first time with the most numerous North Asan people, the Tungus (Evenki). Three major tributaries of the Yenisei were named after them. In 1618, the Cossacks founded the wooden fortress Yeniseisk that became one of the main support bases of the Russian settlements in Siberia, and 10 years later, the Krasny fortress that became Krasnoyarsk. Along the Upper Tunguska (Angara), the Cossacks penetrated into the "land of Brats" (Buryats), where they later founded the town of Bratsk (1631). For the first half of the 17th century, the Cossack detachments discovered, traveling by sea, the mouths of the other East-Sibrian rivers: from the Pyasina to the Kolyma. In the middle of that century, from the north, from the Lena river, the Russians advanced into the Prebaikalia and Transbaikalia, and Kurbat Ivanov in 1643 was the first to reach Lake Baikal in the area of Olkhon Island. In 1661, Yakob Pokhabov founded the Irkutsk fortress.

The Russian advance eastwards continued. As early as 1639, the members of the detachment headed by Ivan Moskvitin reached the Sea of Okhotsk, and within the following 15 years they discovered and described the greater part of its coast. In 1648, the expedition headed by Semyon Dezhnev and Fyodor Popov was the first to traverse the strait between the Arctic Ocean and the Pacific Ocean thus demonstrating that the North American and the Asian continents are not connected. Later, Dezhnev discovered Chukotka Peninsula and Anadyr Bay, traversed the Koryak upland and explored the Anadyr river and the Anadyr lowland.

Approximately in the same years, Yakutsk became the starting point for the discovery of new ands in the south of the Far East, in the Amur basin. The detachments headed by V.Poyarkov, E.Khabarov, P.Beketov, O.Stepanov, and others reached the tributaries of the Amur first, the Argun and Shilka, and then the Amur itself in its middle and lower parts, traversed and described its main tributaries, the Zeya, Ussuri, and others, erected several fortresses, for the first time met with the aboriginal Far-Eastern peoples: the Daurs, Nanai, Nivkh, and others. It is Pyotr Beketov who for the first time surveyed the entire Amur river way, as far as its mouth. First hydrographical schematic maps of the Amur basin were compiled.

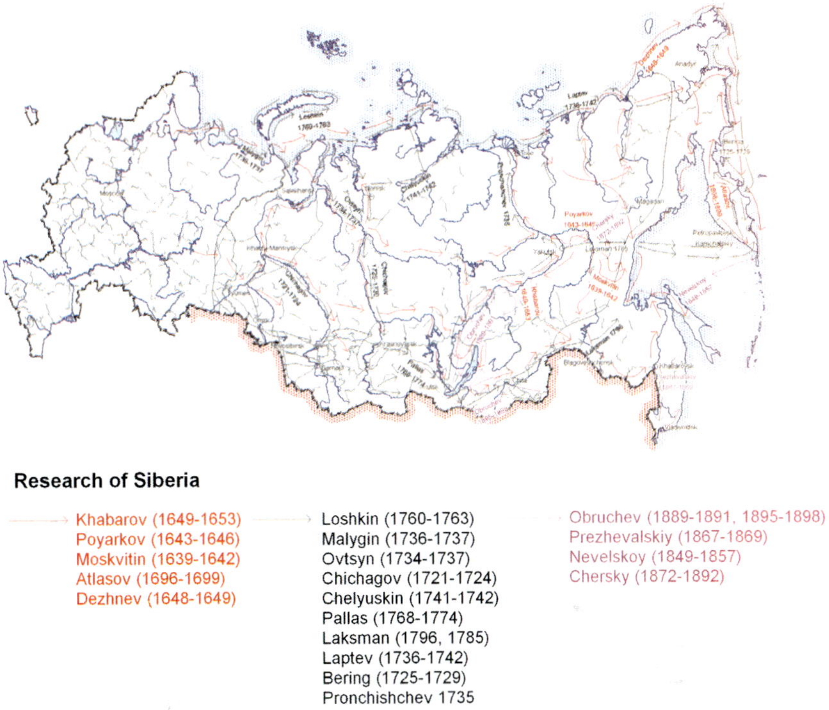

Figure 13: Summary map of Russian campaigns.

Thus, toward the end of the 17th century, actually for a hundred years, the Russian explorers (military and traders), in the most complicated natural conditions, traversed, described and partly annexed to Russia huge North Asian expanses as far as the Pacific Ocean with a total area of more than 10 million square kilometers. That can be called the Russian feat and honor from the era of Great Geographical Discoveries. The geographical result of that stage may well be considered the "Drawing Siberian Land" prepared on the instructions given by the Tobolsk voivode (governor) Pyotr Godunov. The map now shows Lake Baikal, the Amur, and Kamchatka, and the schematic representation reflects the level of knowledge of that time.

The last decades of the 17th century are characterized as the beginning of scientific geographical research on Siberia. It is associated primarily with Semyon Remezov, who surveyed purposefully the Irtysh and Ishim basins; the chief thing, however, is that by the year 1701 he compiled the "Drawing Book about

Siberia", a unique summery of materials concerning the eastern territories, based on the descriptions and schematic maps from the 17th century.

The first Siberian comprehensive expedition should be considered the travel of D.-G.Meserschmidt (1720 – 1727). Alternating routes on land and rivers, he traversed on foot and horseback the entire southern part of West and East Siberia to reach the Transbaikalia, surveyed the valleys of the Ob, Tom, Chulym, Upper and Middle Yenisei, Lower Tunguska, Upper Lena, and Angara. As a result, Messerschmidt was able to gather large mineralogical, botanical-zoological, ethnographical and archeological collections, specified the hydrography of many rivers, was the first to discover such a phenomenon as permafrost. As a result, he produced 10 volumes of the "Survey of Siberia, or Three Tables of Simple Kingdoms of Nature" in the Latin language.

In Siberia, the most interesting and productive were ten-year-long travels undertaken by I.Gmelin visiting all South-Siberian mountain systems, Yakutia, and the Middle Siberian upland. He was the first to single out West and East Siberia, set up the Siberia-first weather stations and prepare five volumes of the "Flora of Siberia".

The outstanding national historian and geographer Vasily Tatishchev carried out his activity during 1730-1750. Not only did he lay down the foundations for the theory of geographical science (providing the definition, classification, and a substantiation for singling out the economic geography, and outlining the linkages between history and geography, etc.), but he was the first (1726) to give a rationale for the creation of the transportation network in Siberia, advocating that it is the Ural Range that is the natural boundary between Europe and Asia.

In 1758, M.V. Lomonosov became head of the Geography Department that was organized as early as 1739. It pursued its activity with a focus on the eastern territories along several main lines. In the first place, it is the substantiation for the idea of the Northern Sea Route, and for setting up several expeditions for studying the Arctic. Secondly, it is the generation of maps on a constant basis, including the first maps on economic geography of Asian Russia. It is the judgments advocated by Mikhail Lomonosov regarding the causes of the continental climate in Siberia, who correctly explained the more severe colds at Irkutsk and Nerchinsk when compared with Tobolsk, by the different distance from the seas. Finally, it is the launching of the journal by the Academy of Sciences in which geographical topics occupied the central place; more specifically, the well-known works by G.Miller about the Arctic and the Amur basin were published in that journal.

In the last third of the 18th century, of important significance were the Academy's Siberian expeditions headed by Pallas and Georgi. P.-S.Pallas' travel lasted from 1768 to 1773 reaching southern Siberia and the Transbaikalia. It provided a large number of valuable comprehensive descriptions and surveys to shortly appear in three volumes, and they have retained their value to date. I.G.Georgi, after his expedition to Lake Baikal in 1772-1774, provided a first scientific description of the great lake, its shores and inhabitants and suggested that it has a catastrophic origin.

An important stage of Siberian research should be considered the expeditions headed by Alexander F.Middendorf (1843-1845). As a result, he was the first to determine the southern boundary of permafrost in East Siberia, investigate Taimyr, develop a classification of the tundras, and to provide proofs of the zonal distribution of vegetation on this territory, and a characterization of its climate.

In the 19th century, the prime object became a systematization and summarization of the materials obtained, as well as taking advantage of the new capabilities of cartographic and instrumental methods. That activity was almost entirely concentrated in the Russian Geographical Society (RGS) that was created in 1845, and its regional divisions: the Siberian Division at Irkutsk (founded in 1851, renamed the East-Siberian Division in 1877), the West-Siberian Division at Omsk (1877), the Priamursky Division at Khabarovsk (1894), the Yakutsk Division (1913), and their respective commissions, subdivisions, and branches.

The principal method of obtaining geographical knowledge was through expeditions, as before. It should be noted that most expeditions, unlike the preceding stage, were comprehensive geological-topographical-geographical-ethnographical expeditions and were accompanied by high precision astronomical, geodetic and meteorological research; hence their materials have a special value.

The end of the 1860s saw the opening of the "Polish" chapter of research. Polish exiles became leading scientists of the Siberian (later East-Siberian) Division of the RGS. B.Dybovsky and V.Godlevsky did research at and on Baikal, in the Baikal region, and in the Amur basin. A.Chekanovsky investigated the Irkutsk province, and the Middle Siberian upland. A particularly important contribution was made by Ivan Chersky, who investigated the Eastern Sayan mountains, the Kuznetsk Alatau, the mountains of the Prebaikalia, and Baikal's shores. At the end of his life career, he surveyed the mountains of northeastern Asia.

That period of time was characterized by the appearance of analytical surveys and reviews as a result of almost all expeditions in numerous specialized publications of the RGS and its regional subdivisions which have retained their value to date. Besides, a number of classic works were published thanks to the long-term President of the RGS Pyotr Semenov: his translation from the German of the "Land Ownerships of Asia" by K.Ritter, the Geographical-Statistical Dictionary of the Russian Empire, a series of books on regional studies "Russia", and the "History of the 50 Years of Activity of the Imperial RGS. 1845-1895".

In Soviet times there was a cardinal change of the objectives, organization and methods of geographical studies of Asian Russia. The objectives then were to secure the industrial development of the eastern regions and the military-political priorities. The Academy of Sciences gradually resumes the work on organization of research, and in 1938 the RGS is in charge of it. In Moscow and

Leningrad appear specialized geographical institutions, both integrated and specializing in branches of geographical knowledge, because geography, as science, was becoming strongly differentiated. University-based geographical research was gaining momentum; this was especially true for Moscow and Leningrad Universities. For the eastern regions, an important role was played by such governmental geographical-economic institutions as the Committee for the North, the Council on the Development of Productive Forces (CDPF), the Commission for the Development of Natural Productive Forces, and others. The importance of accurate geophysical, hydrometeorological and aerial photography precise methods was enhanced, and a network of specialized observation sites and research stations were set up, including on the territory of Siberia.

The integrated expeditions during the 1940s-1950s in the south of East Siberia had a special character; they were organized in order to study the hydropower potential and to create, on its vasis, powerful territorial-production complexes (TPC). The closer dependency of the planned complexes (based on utilizing local hydropower, water, forest and mineral resources) on natural conditions and the geographical (including economic-geographical) and, especially, transportation-geographical location, was the driving force to enhance the geographical component in the investigations that provided plans and design documentation. The challenging problems of creating the East-Siberian HEPS, industrial centers, and TPC were discussed on-site through a series of scientific conferences on the development of productive forces of Siberia, the first of which was held at Irkutsk in 1947.

The expeditions and conferences played a large part in the development of the eastern regions' economic potential. However, there was a serious drawback if research was being done by scientists from Moscow only. That resulted in the creation of institutions in Siberia, the first of which were the Limnological Station in the settlement of Listvyanka at Lake Baikal (since 1925), the Yakutsk Permafrost research Station at Yakutsk (since 1941), and the Economics and Geography Division at Irkutsk (since 1949). Geography sub-faculties and faculties were also functioning in the universities of Tomsk, Irkutsk and Vladivostok. Actually, however, the development of a network of geographical institutions in the country's east started only after the creation of the Siberian Branch of the USSR AS in 1957.

In addition to the Institute of Geography SB RAS (in 2005 it was named after academician V.B.Sochava, its first director. His scientific work is discussed below), geographical research in the Siberian Branch RAS is also pursued by a number of institutes, having divisions and laboratories specializing in geography: the Limnological Institute (Irkutsk), the Institute of Water and Ecological Problems (Barnaul), the Baikal Institute of Nature Management (Ulan-Ude), the Institute of Permafrost Studies (Yakutsk), the Institute of the Earth's Cryosphere (Tymen), the Institute of Monitoring of Climatic and Ecological Systems (Tomsk), and the Institute of Natural Resources, Ecology and Cryology

(Chita). Scientists from institutes specializing in geological, biological and economics research take an active part in these research efforts.

Specialists in geography are trained (with concurrent related research) by classical and teachers' training universities of Tomsk, Irkutsk, Vladivostok, Barnaul, Tyumen, Krasnoyarsk, Nizhnevartovsk, Chita, and other cities of Asian Russia. Here there also are geographical and geographical-ecological subdivisions of a number of ministries and agencies (natural resources, transport, hydrometeorology and environmental monitoring, etc.) which are doing a great deal of multifaceted research in the field of nature management, in close association with geographical research. Special mention should be made of the long-term research within a network of specially protected territories in all eastern regions of the country.

As a result of pooling efforts made by the above institutions, with the involvement (especially at an early stage) of some scientists from the country's central regions, and geographical studies of Asian Russia for the last 50 years, an impressive series of important theoretical generatalizations was created. Integrated expeditions have been carried out in connection with the projects of industrial centers, transportation systems, and economic development of the eastern expanses: the projects of the Bratsk-Ust-Ilimsk and Nizhnevartovsk TPC, BAM (Baikal-Amur Railroad), the transfer of the Siberian rivers to Central Asia, pipelines from East Siberia to the Pacific, and others.

INVESTIGATIONS OF THE V.B. SOCHAVA INSTITUTE OF GEOGRAPHY SB RAS

The leading geographical institution of Siberia, the V.B.Sochava Institute of Geography SB RAS, was established in 1957 at Irkutsk, simultaneously with most institutes of the Irkutsk Scientific Center. Its major research areas include the studies of natural resources, industrial production, transport, and agriculture of Siberia and the Far East for the purposes of their optimal distribution and development. Of significant importance has also been and is the cooperation with the academies of sciences of Asian countries, primarily Mongolia.

From the very beginning of the Institute's research activities there was a need for fundamentally novel theoretical concepts, meeting the challenges of the time and reflecting the specific character of geographical research on the new territories under economic development. These challenges were embodied in the doctrine of geosystems created by an outstanding national geographer, Institute director, academician V.B. Sochava, who described it in his monograph titled "Introduction to the Doctrine of Geosystems" [7]. Geosystems represent a special class of open, hierarchically organized dynamical systems. The hierarchic structure of the geographical envelope makes it possible to consider systems of all levels – from planetary to local – as a single whole, combined into a unified natural organization, with linkages and development patterns inherent in each level. Such an approach makes it possible to combine a host of concepts into a geographical space of state variable of a geosystems in the form of ordered series.

Assessments of the evolutionary-dynamical state of geosystems are of exceptionally important significance. Short period pulsations of the natural regimes that do not alter the underlying, structure-forming properties, create a variety of dynamical properties of individual natural components. To investigate these properties, comprehensive research efforts were undertaken at permanent stations under different natural conditions [1]. Based on the results of those research efforts, and on the modeling methodology, geographers were able to suggest a productive idea of cooperation between man and Nature and, hence, an optimization of this interaction. Besides, any dynamical manifestations in natural regimes carry indicators of evolutionary transformations, which is also of much scientific interest.

V.B. Sochava identified, in addition to full geosystems, partial geosystems. They represent a system in which one of the natural components as a system-forming element is adjusted to the entire set of factors influencing it. According to typological characteristics and spatial dimensions, these connections have in each particular case their specific character, are spatially localized and can be combined, by the principle of adjacency, with other partial systems thus creating a next hierarchical level of particular spatial ordering. On this basis, the Institute has been and is successfully pursuing landscape-geochemical, hydroclimatic, glaciological and biogeographical research. Yet, the unifying idea implies achieving a geographical synthesis, leaning upon a knowledge of the patterns inherent in individual natural environments as well as in geosystems in general.

A knowledge of the structural and functioning patterns of geosystems is also necessary when developing cartographic documents. Cartographic methods, in turn, are a powerful tool for gaining insight into spatial organization, primarily spatial structures. Thematic mapping has headed the list of the Institute's research projects since its establishment. The Institute generated maps as well as special series of maps both for the whole of Asian Russia and for its parts, reflecting separate sectoral branches, such as vegetation maps, and soil cover maps. Atlas mapping efforts were successfully undertaken. A first impressive product was the Atlas of the Transbaikalia (1967) that embodied the bulk of knowledge concerning the natural environment, economy and population of the region available then. The most important work was the National Atlas of Mongolia (1990) which combines into a common ideology of systemic mapping 276 maps at a general scale 1:3000000. This work, the result of a comprehensive inventory of the natural conditions and resources, the characteristics of the sectoral, social and territorial structures of the economy, population problems, and environmental conditions, has become for many years ahead the key information base for the development of our neighboring Asian country. The Institute produced the atlases of the Amur region, the territory of the Kansk-Achinsk fuel and energy complex, and of Lake Hovsgol (Mongolia).

The highest level of fundamental mapping was attained by the creation of the Aatlas "Irkutsk Region: Ecological Conditions of Development" (2004), having no parallels worldwide and nation-wide. It represents a systemic collection of maps, reflecting the ecological conditions of the region's balanced development and is, at the same time, a mapping brand of the Irkutsk region. The atlas has assembled 166 maps, complemented with diagrams, plots, schematic representations, tables, photographic maps, and photographic scenes. The maps are accompanied by scientific explanatory notes, containing the principles and methods of map compilation, a brief analysis of their contents, and the method of utilizing the information presented.

The organization of the Atlas implies a comprehensive portrayal of the natural and anthropogenic factors that are responsible for the shaping of the region's ecological situation plus the natural-ecological framework of formation of the territory's socio-economic profile. The Atlas details the natural and socio-economic formation conditions of the ecological situation. The main section in the Atlas deals with the anthropogenic factors that are responsible for the ecological situation, and the environmental quality. It provides a comprehensive, integrated portrayal of the factors of anthropogenic modification of environment as well as the consequences of this interaction, such as the effects of transformation of biotic elements, water bodies, landscapes, etc. These representations are synthesized by the maps of ecological zoning and regionalization, and networks of specially protected territories. Most of the methods for developing classifications and for their cartographic interpretation were specially created for this Atlas and have no equal worldwide. In addition to the "hardcopy" of the Atlas, its electronic version is also available. A recognition of the important significance of the Atlas was the fact that the team of cartographers was awarded the State Premium of the Irkutsk region for the year 2006.

Traditionally, much attention is paid by the Institute to the population issues. Originally, Siberia was populated by small tribes and clans; subsequently, they were assimilated by the more numerous ethnic groups (the Yakuts, Buryats, and Khakas), and by Russians. Aboriginal ethnic groups had played a large part for the shaping of the contemporaneous population; although after the annexation to Russia their population increased by a factor of 8 or 9, their share in the total population did reduce (from 30% in the 19th century to 10% at present). The Irkutsk school of specialists in the field of population geography under the guidance of academician V.V. Vorobiev formulated a number of original ideas concerning the shaping of the population in the country's eastern regions. Specifically, it was demonstrated that no mass "extinction of native peoples" never occurred in Siberia. On the other hand, as indicated above, there is an obvious tendency toward a decrease in the population due to westward emigration, and to an abrupt decrease of natural growth rates. Particular attention has been paid recently to the problems of ethnocultural interaction.

The Siberian school of social and economic geography has been functioning at the Institute for many years. It is concerned with issues relating to natural resource assessments, the study of the specialization and comprehensive development of the Siberian regions, the problems of the eastward shift in the distribution of productive forces, a characterization of the territorial-production complexes, and the prospects of their development. As a result of the long-term research, it was possible to create and pursue such research areas as structural regional studies and regional nature management, medical geography, geographical expert examination, resource management, and others.

During the period of cardinal restructuring of Russia's economy, quite varied processes were taking place throughout Siberia, which have drastically changed its socio-economic profile. Economists-geographers of the Institute are currently engaged with the studies of these tendencies and their consequences. For instance, farm enterprises and their socio-economic stratification are emerging. To identify the chief tendencies of this process, it is necessary to determine the characteristics of the adapting of this emergent agricultural sector to the new conditions of the food market, and to define its prospects in the regional industrial complex. Furthermore, a particularly important role is played by the farm sector as a consolidating core of the labor potential of the rural areas.

Equally important is the development of new approaches to assessing the competitiveness of the territory through an analysis of its consumer potentials, and the diversification level of the internal market as the initial stage of sustained self-development. Market stimuli lead to a new spatial structuring of the productions sectors, and to the creation of new commodity markets. This is particularly clearly manifested in the formation of the consumer environment at a lower administrative level, municipal formations.

It cannot be too highly stressed that geography as a powerful tool for addressing practical issues has special social implications. The national applied geography has its origin in Irkutsk and has filled a highly important place in the Institute's research efforts. As is known, it is geography (the only science to date) that is able to integrate natural sciences and the humanities to gain a better understanding of the environment. To translate a fundamental knowledge to the resolution of applied problems requires a special target-oriented comprehension and review of our concepts not only with regards to the natural part of the environment but also the entire set of interacting (within separate territories) social, political, economic and other spheres. Target-oriented integration of the sectoral research areas is a highly complicated and multifaceted process. It must be built upon very different tools: informational, planning, legal, etc. In any case, however, they must be target-oriented.

A most complicated vector of development of applied geography must involve planning the territorial (spatial) development. Quite varied analytical means and procedural methods are invoked to provide a broad interpretation of this research area. At any rate, however, planning is a poly-systemic iteration process, based on the general concepts of the principal formation patterns of the territory's dynamical and structural characteristics. The development of tools for integrated management of the development through territorial (spatial) planning at the V.B.Sochava Institute of Geography SB RAS has been carried on during many years. To date this effort has crystallized into the methodology of so-called landscape planning that has assembled all body of knowledge of a model territory.

In 2007, the V.B.Sochava Institute of Geography SB RAS celebrates its 50th anniversary. For the years since its establishment it has had to cope with a host of challenges to become a major geographical center in Russia's east. At present its staff are more than 200 employees, including 27 doctors of science. They represent almost all geographical specialties: landscape scientists and hydrologists, soil scientists and botanists, and cartographers and economists-geographers, who are able to resolve any complicated problems. The Institute has 10 laboratories and 2 doctoral dissertation councils in 5 geographical specialties; every year the Institute publishes 15 – 20 monographs and produces tens of maps of a different purpose; since the year 1980, the Institute has been publishing the journal "Geography and Natural Resources". The Institute maintains relations with geographical institutions and geographers of various countries: Germany, Mongolia, China, and others. The main results of the Institute's scientific activity are summarized in this collection of papers.

CONCLUSION

Siberian geographers are facing grandiose tasks. In the first place, it is comprehensive and hence geographical assessment of the place of Siberia and Asian Russia in general, in the global development. The region's giant resources and their low development, the unique natural environment and its chaotic modification as a result of the economic activities, and the up-to-date scientific potential and its inadequate utilization call for fundamentally new views of the development strategy within the context of balanced interests and rationality in nature management. The research objectives over the past several years have been to gain a better understanding of the specific character of natural processes under the circumstances of global and regional, climatic and man-induced changes, to study the territorial organization of society in the new socio-economic conditions, and to define the geographical aspects of integration of the eastern regions' economy into the world economy, primarily the Asian economy, with the ecological priorities unaltered. It is the geographical science that is able to meet the challenges of the present time by creating the prerequisites to the conflict-free successful development of such a grandiose and unique natural phenomenon as Siberia, for the benefit of Russia and all mankind.

REFERENCES

1. Mikhailov N.I. Siberia. Moscow, 1956, 382 p.

2. Mikhailov N.I. Nature of Siberia. Moscow: Mysl, 1976, 159 p.
3. Duchkov A.D. and Balobayev V.T. Evolution of the Thermal and Phase State of Siberia's Permafrost Zone. Global Changes of the Natural Environment. Novosibirsk: Publishing House of SB RAS, "Geo" Branch, 2001, pp.79-104.
4. Rakovskaya E.M. and Davydova M.I. Physical Geography of Russia. Pt. 2. Moscow, 2001, 304 p.
5. Gvozdetsky N.A. and Mikhailov N.I. Physical Geography of the USSR. Moscow: Mysl, 1978, 512 p.
6. Parmuzin Yu.P. The Taiga of the USSR. Moscow: Mysl, 1985, 303 p.
7. Sochava V.B. Introduction to the Doctrine of Geosystems. Novosibirsk: Nauka, 1978, 320 p.

PART II

PHYSICAL GEOGRAPHY

CHAPTER

2

Geomorphological Research on Siberia

L.N. Ivanovsky and V.B. Vyrkin

The creation of a scientific school of exogenous geomorphology in Irkutsk is closely related to the general direction of geomorphological investigations pursued at the V.B. Sochava Institute of Geography SB RAS, as outlined in the late 50s-the 60s by academicians I.P. Gerasimov and V.B. Sochava, and professor S.S. Voskresensky, as well as to the sector of dynamic geomorphology organized at the institute in 1965. Since the establishment of the sector, geomorphologists were taking an active part in comprehensive geographical research at the Institute's stations and during expeditions to various regions of Siberia.

The stations were set up in the steppes of the southeastern Trans-Baikalia, in the taiga downstream from the Angara river, on the plains of West Siberia, at the foothills of the Western Sayan mountains, and within mountain troughs in southern Siberia. Semi-station-based observations were also made on the slopes of the Eastern Sayan mountains, in Altai, Kodar, Udokan and other mountain ranges in southern Siberia. The principal goal of station-based work was to study the dynamics of exogenous processes and the development of landforms at a topologic spatial level, i.e. to identify the relief-forming mechanisms.

Stationary observations have helped to find out that the water-erosion-driven intensity of substance ablation and accumulation on the steppe slopes in south-eastern Zabaikalye is linked to short-period hydroclimatic rhythms. This dependency is disturbed on slopes affected by human economic activities. These rhythms also determine processes of stronger or weaker wind erosion.

Processes of mass substance slide (solifluction, defluction, and the thermo-, hydro- and cryogenic varieties of deserption) were studied at the foothills of the

Western Sayan and at the upper forest border, in the taiga in the Lower Angara region, on the mountain slopes in the Baikal region, and in the Southern-Minusinsk and Nazarovsk depressions, their intensity was determined, as well as immediate successions of one process by another were revealed [1, 2]. Major firn-melting and nivation process development patterns were described for the mountains in the Baikal region and the northern Trans-Baikalia [3, 4]. The examination of current exogenous relief-forming processes has revealed their domination by hydrogenous slide and flow on mountain-taiga slopes, and by cryogenic on-slope and gravitation on-slope processes, above the upper forest border.

The occurrence and morphology of both active and stabilized gullies were studied in much detail, issues of their genesis discussed, and efforts to predict their development made. With abundant showers in loess-type sediments, even one shower was found to be sufficient for formation of hundreds-of-meters-long gullies.. The density of gullies was estimated for the Southern-Minusinsk depression, and in the Trans-Baikalia with the result that that the potential gully-formation is determined by physical-geographic conditions, while its duration is governed by economic development. The gully-formation potential and the agricultural land-use influence the distribution and density of gullies. An effort was made to develop a morphological classification of gullies, with their identification based on characteristics of landforms and loose sediments. Forms of gully development shapes were identified, as well as processes of their development and the stages of their transformation into inactive gullies and valleys [5–7].

Fundamental patterns of temporal and spatial changes of aeolian and water-erosion processes were discovered found for the steppe slopes. Quantitative models were used to assess estimate the Nazarovo depression trough in terms of potential wash-off along the slopes. Erosion indices were calculated for rainfall over the territory of Khakasia, the erosion potential of relief and agrocoenoses were studied, and current exogenous processes on agricultural lands were mapped.

At a station located on the right bank of the Irtysh river close to the mouth of the Missiya river, I.B. Petrov [8] examined riverbed landforms: shoals, rifts and islands, wavy sand bars, beaches; progression of bank destruction,, and the types and rates; flood-plain, its types and development; the morphology of flood-plain terraces and tributary valleys; and compiled a predictive map for the evolution of the banks of the Irtysh river downstream.

Analyses of data from stations led to the conclusion that it is necessary to devise a special direction within geomorphology: the study of smaller landforms, or topologic geomorphology [9], supported by its own methodology and research tools, and extensively pursued at the V.B. Sochava Institute of Geography SB RAS until the 1990s.

The 1970s saw the stage of summarizing research results on the history of relief evolution in Siberia and the Far East, reflecting L.N. Ivanovsky's efforts to tackle the problems of Quaternary glaciation [10] and river-terrace formation in mountains [11].

In 1981, L.N. Ivanovsky [12] published a monograph titled "The glacial Geomorphology of Mountains" which was highly acclaimed by the geographic community; he was awarded a Przevalsky gold medal by the Russian Geographic Society.

By the early 1980s, it became clear that to continue the research into exogenous relief-formation requires elaborating the relevant theoretical methodological framework. In this connection, V.B. Vyrkin [13] developed a classification of exogenous processes, based on Shantser's classification of genetic types of continental sediments [14], and on the inclusion of primary stages and agents of the transport of materials across land surfaces. To comply with the principal rules of subdividing the body of the concepts, it was proposed to categorize the exogenous processes as types, subtypes, classes, groups and elementary processes in a classification of exogenous processes. Based on this classification, principles and methods for mapping current exogenous processes in relief-formation were proposed and tested [15], and medium- and small-scale maps for the depressions troughs in the Baikal region and Trans-Baikalia were generated, as well as 1:2,500,000- and 1:1,000,000-scale maps for current exogenous relief-forming processes in the Irkutsk and Chita oblasts, and in the Republic of Buryatia.

The station-based and regional geomorphologic investigations to date have also dictated the need to distinguish, among a host of processes observed on the territory, the leaders, without which no exogenous landform can be produced. The leading process must be identified from taxa, given the descending lithodynamic substance flow [16–18].

By identifying the leading processes, it was possible to determine structures of the processes for any study territory. For Lake Baikal-type depressions it was shown that the spatial structure of current exogenous relief-forming processes, shaped by a combination of leading processes interacting across space, is formed and acquires its own individuality within the humid moderate morphoclimatic zone, depending on the morphology of the depressions, the result of their historic development during the Anthropogene [19]. Analysis of spatial structures of processes was later complemented by identification of temporal structures of processes (chronostructures) [20]. Work still continues and forms a methodological basis for implementation of a research project titled "Chronostructures of landscape-forming processes at the time of global and regional changes in southern Siberia's natural environment".

The development of concepts of descending lithodynamical flow, identification of the leading process within the exogenous relief-formation, and analysis of the spatio-temporal structure of the process form a basis for studying the processes in southern Siberia at a regional, local and topologic levels. Identification of the leading exogenous processes and analysis of their structures are made difficult because of a continual succession of processes and their structures across time and space. The successions of the processes can be of different types and have different values: seasonal, annual, or corresponding to short or somewhat longer hydroclimatic rhythms, as well as to rough topographic surfaces. Hence there emerged a (not yet realized) idea of producing "on-the-spot maps" of exogenous processes based on the hydroclimatic rhythm at any time scale. Observations at stations show that successions of on-slope water-erosion and aeolian processes are consistently evolving at short hydroclimatic rhythms, and areal and linear erosion (gullies) on fields give way to wind erosion [21–22].

Each smaller landform that is influenced by the leading process can grow in size to a certain limit; after that, however, this process is reversed to change the morphology of topography, with the production of a new landform overlapping the remaining landform of a different genesis.

A number of forms succeeding each other were studied in conditions of alpine-type topography in southern Siberia, starting with a very simple nival niche on the slope along the original erosion groove. The nival niche, because of its increasing volume and snow-accumulation, turns into a nival corrie. The nival corrie expands, accumulating an ever increasing amount of snow that is unable to melt over the course of a warm season. The accumulated snow transforms into ice, and the leading process reverses to form a glacial corrie. The expanding corrie is gradually destroyed to produce horn-peaks, giving rise to an equiplain [10].

When explaining the formation of alpine landforms in Siberian mountains, L.N. Ivanovsky [17] was the first to introduce to geomorphology such concepts as 'parageny'*, their associations, and 'paragenesis' of forms and processes. By parageny is meant a neighborhood of landforms, while paragenesis embraces the phenomena occurring between these forms (exogenous processes). Paragenesis in geomorphology comprises 2 systems of evolving landforms which are in parageny. One produces a certain chain of processes transferring material from one object to another thus creating a paragenetic series, while the second is not characterized by the transfer of material, but by the interaction of the forms taking place at the junction of the slopes.

Parageny and paragenesis were examined for rapid mass destruction of younger alpine and medium-height mountain landforms, as well as for areas with rocky, sharp watershed ridges of various orders of altitude, a certain regularity was detected in the character of destruction of larger alpine-type relief forms, e.g., corries, replaced by large denudation craters, avalanche chutes, new nival niches, and smaller craters on the slopes of a different height, which, developing in parallel, contribute to smoothing alpine topography. Such a course of destruction of the formerly alpine landforms is characteristic of rejuvenated Siberian mountains. The development of alpine landforms in these mountains is taking place though a consistent replacement of smaller forms by larger ones and, vice versa, dividing larger forms into a number of smaller ones.

In summarizing, it can be said that the past several decades of scientific research done by geomorphologists at the V.B. Sochava Institute of Geography SB RAS were marked by a successful development of a new direction in exogenous geomorphology, involving a set of original theoretical-methodological approaches to studying current exogenous processes of relief-formation, capable of being laid down as corner-stones in the development of the exogenous morphogenesis theory. They include elaborating problems of unveiling leading geomorphologic processes, their mapping, analyses of spatial and temporal structures of current exogenous relief-formation, parageny, and paragenesis, as well as process succession development. This direction, conceived primarily based on developing stationary research, is currently being enriched by means of new methodological techniques used in the analysis of relief-formation processes, as reflected in major publications by a team of geomorphologists at the V.B. Sochava Institute of Geography SB RAS. Thus, the nearly 40 past years have witnessed the creation at the Institute of a now-active scientific school of exogenous morphology which is recognized by this country's geomorphologists, has acquired a profile of its own and holds much promise as regards its future.

REFERENCES

1. Vyrkin V.B. Boulders on naked mountains in the Baikal region and the Northern Transbaikalia. Current Relief-Formation Processes in Siberia. Irkutsk: IG SB RAS Publisher, 1978, pp. 88-108.
2. Bazhenova O.I. Loose material movement patterns on forest-steppe slopes in the Nazarovo depression. Geografiya i prirod. Resursy, 1982, No. 2, pp. 98-103.
3. Vyrkin V.B. Physical weathering of rock and microclimate in firn-basins in the Baikal region's and Northern Trans-Baikalia's mountains // Geografiya i prirod. Resursy, 1980, No. 4, pp. 101-107.
4. Vyrkin V.B. Firn-basins in Pribaikalye and the Northern Trans-Baikalia's mountains. Geografiya i prirod. Resursy, 200, No. 4, pp. 90-95.
5. Lyubtsova E.M. Smaller erosion relief-forms in the steppes of the Trans-Baikalia. Integrated Geographic Research of Frontier Regions in Siberia. Irkutsk: IG SB RAS Publisher, 1980, pp. 14-23.
6. Ivanovsky L.N., Titova Z.A., and Vyrkin V.B. The problem of predicting current exogenous relief-formation processes. Geografiya i prirod. Resursy, 1983, No. 4, pp. 20-25.
7. Ryzhov Yu.V. Gullyingin southern Siberia. Geografiya i prirod. Resursy, 1995, No. 3, pp. 101-110.
8. Petrov I.B. The Ob-Irtysh Flood-Plain (Typization and Qualitative Land Assessment). Novosibirsk: Nauka,1979, 136 p.
9. Ivanovsky L.N. Aspects of geotopology in relief studies. Topologic Aspects of the Geosystems Doctrine. – Novosibirsk: Nauka, 1974, pp. 214-240.
10. Ivanovsky L.N. Mountain glaciation and relief. In: Problems of Exogenous Relief-Formation. Moscow: Nauka, 1976, Vol. 1, pp. 90-188.
11. Ivanovsky L.N. River Terraces of Mountain Rivers and Their Longitudinal Profiles. Problems of Exogenous Relief-Formation. Moscow: Nauka, 1976. Kn. 2, pp. 248-262.
12. Ivanovsky L.N. Glacial Geomorphology of Mountains. Novosibirsk: Nauka, 1981, 173 p.
13. Vyrkin V.B. Classification of exogenous processes of land relief-formation. Geografiya i prirod. Resursy, 1986, No. 4, pp. 20-24.

14. Shantser E.V. Essays on the Doctrine of Genetic Types of Continental Sedimentation. – Moscow: Nauka, 1966, 239 p.
15. Vyrkin V.B. Basic principles of mapping current exogenous processes in relief-formation Geografiya i prirod. Resursy, 1991, No. 2, pp. 163-167.
16. Ivanovsky L.N. Exogenous Lythodynamics of Mountainous Countries. Novosibirsk: Nauka,1993, 160 p.
17. Ivanovsky L.N. Parageny and Paragenesis of Mountain Topography in Southern Siberia. Irkutsk: IG SB RAS Publisher, 2001. – 142 p.
18. Vyrkin V.B. Current Exogenous Relief-Formation of Baikal-Type Depressions. Irkutsk: IG SB RAS Publisher, 1998, 175 p.
19. Vyrkin V.B. The structure of current exogenous relief-formation in Baikal-type depressions. Geografiya i prirod. Resursy, 1996, No. 1., pp. 13-23.
20. Vyrkin V.B. Chronostructure of plane denudation processes on slopes. Geografiya i prirod. Resursy, 1992, No. 4, pp. 146-151.
21. Vyrkin V.B. Geomorphologic aspects of erosion on agricultural lands in the Baikal region's depressions. Geografiya i prirod. resursy., 1991, No. 1, pp. 38-44.
22. Bazhenova O.I., Lyubtsova E.M., Ryzhov Yu.V., and Makarov S.A. Spatio-Temporal Analyses of the Erosion Process Dynamics in Southern East Siberia. Novosibirsk: Nauka, 1997, 208 p.

CHAPTER

3

Trends of Exogenous Relief-Formation Regimes in Southern Siberia in the Second Half of the 20th Century

O.I. BAZHENOVA

Interest in the study of exogenous relief-formation processes has quickened worldwide during the last several decades, which is largely explained by the shift of traditional geomorphological studies to addressing environmental problems in which the intensity of exogenous processes is a key indicator of the state of environment. In this context, the attention paid by the Geomorphological Commission of the International Geographic Union to the study of the influence of a global warming on the course of geomorphological processes, and to assessments of their contribution to environmental change becomes understandable [1]. It is necessary to know how the intensity and direction of these processes are changing, as well as the functioning regime of geomorphological systems in the case of fluctuations of climatic parameters, i.e. the predictive focus of geomorphological research comes to the forefront, the methodological framework of which is till poorly elaborated.

The principal methods of geomorphological forecasting include the technique for extrapolating the relief-formation trends [2], which can be implemented only when time series of indicators of the intensity of processes and factors governing them are available. Such series were obtained from studying the modern morphoclimatic regimes of a zone of island steppes and forest-steppes in southern Siberia [3] to be sued in identifying trends of exogenous relief-formation that emerged in the second half of the 20th century in connection with climate change. The emphasis was on investigating the response to the climatic signal of the erosion, aeolian and cryogenic processes, playing a crucial role in exogenous relief-formation. A number of serious ecological problems are associated with the occurrence of these processes.

In the case of a predictive assessment, a large number of theoretical and empirical models are used to select dependencies in which the climatic parameters characterizing the behavior of the processes, are supported by time series. The validity of using them in the region under consideration is confirmed by experimental data from permanent observation stations [4]. The correlation coefficients of the measured and calculated rates of the processes are rather high. The influence of some natural factors on the dynamics of the processes, such as the rock lithology, and the geometry of the slopes, was taken to be invariable across time (for a period of 30-40 years).

To estimate the long-term variability in storm water wash, use was made of the universal erosion equation that was modified at Moscow University [5], where the climatic parameter is represented by the erosion index of storm rain precipitation, the sum of the products of the kinetic energy of rains and their maximum intensity. In order to infer this index, pluviograms for rains with a layer of 10 mm at least from 63 weather stations in Southern Siberia were processed for the last 35-65 years. Wash from the meltwater runoff was determined using the technique suggested by N.N. Bobrovitskaya [6] by analyzing long-term series of water reserves in the snow. The long-term dynamics of the modulus of suspended sediment flow was used as an integral indicator of intensity variation of erosion processes over time. This method is supported by instrumental observations and is recognized by most national and international researchers as the most accurate and unbiased technique [7].

Assessments of the trends of the aeolian processes used the data on the number and duration of dust storms and dust ground wind involved calculating the integral index of deflation by the formula suggested by Wilson and Cooke [8] taking into account the wind velocity, atmospheric precipitation, and air temperatures. Given the particularly important role of the wind activity in the development of aeolian processes, an additional analysis was made of the yearly mean wind velocity for the entire period of instrumental observation. In assessing the cryogenic processes, a study was made of the dynamics of the seasonal freezing-thawing depth and rate, the temperature variation of the soil surface and upper layers, and the number of days with a cold at different depths of an active layer [9], and the probable variations of the velocity of slow mass displacements of earth materials were calculated by the formula suggested by Devdariani [10].

To identify the intensity tendencies of the processes, linear trends of the parameters under consideration were determined by the method of least squares. It was possible to identify three types of trends, reflecting the geomorphological essence of the phenomena: positive (ascending – the rate of the process increases), negative (descending – the involvement of the process in relief-formation decreases), and relatively steady-state (the rate of the process remains unchanged, i.e. at the preceding level). The criterion suggested by Gusarov [11] was used in assigning the trend to a particular type.

A long-term series was considered relatively steady-state if the difference (ΔT) of the mean values (norms) of the parameter between the first (early, T_1) and second (late, T_2) halves of the series being analyzed varied from 0 to ±5 %. When the difference was larger (+5%) or smaller (-5%), the series was considered to have a descending or ascending trend depending on the sign of difference, i.e. according to [11]:

1) relatively steady-state trend ΔT $-5\ \% \geq (T_1 - T_2)/T_1 \cdot 100\ \% \leq +5\ \%$;
2) ascending trend ΔT: $(T_1 - T_2)/T_1 \cdot 100\ \% < -5\ \%$; and
3) descending trend ΔT: $(T_1 - T_2)/T_1 \cdot 100\ \% > +5\ \%$.

In most cases, determination of the trend of a particular exogenous process used time series from several stations; usually, for several parameters rather for a single one. Thus, the conclusion about the decrease in active aeolian relief-formation in the western part of the study zone of steppes and forest-steppes for the second half of the 20th century was drawn on the basis of analyzing negative trends of dust storms, wind activity, and of the integral indicator of deflation using the data from nine stations. Some of these series are provided in Fig. 1.

Our identified tendencies toward a decrease in aeolian activity are also supported morphologically. Thus, in the Barguzin, Southern-Minusinsk and Ubsunar depressions the areas of drist sands are reducing, and the aeolian landforms are becoming overgrowth with vegetation. The attenuation of the aeolian processes in the Minusinsk depressions is associated, in our opinion, not only with current climatic trends, because some positive role here has played by the extensive introduction of the anti-deflation soil-protective system of agriculture. In the Selenga-Khilok steppe and in the dry Udinskaya steppe, the well-established level of dynamics of aeolian processes has persisted to date. An increase in aeolian activity, with an ever increasing occurrence rate of extreme aeolian events, is characteristic for the extra-continental steppes of Central Asia (the Onon-Argun and Kyzyl steppes), and for the Angara-Lena forest-steppe. Overall, the southern part of Siberia shows an expansion of the region of aeolian relief-formation at the sacrifice of forest-steppe areas.

The degradation processes of permafrost in sub-arid landscapes of southern Siberia were considered in detail in [12]. Their indicator is provided by the numerous newly-emerged thermokarst pits along the valleys of the Unga, Zalari, Oka, Kitoi, Irkut, Typta, Bayandaika and other rivers (Angara region), and along the shallow valleys of the Transbaikalia.

Let us now consider at greater detail the character of variation in erosion activity of water streams. A statistical analysis of long-term series of observations of the suspended sediment flows identified watershed basins that are combined into four areas with different behavior tendencies of the erosion processes. The predominant tendency in the first Ob-Yenisei region with a positive trend of atmospheric humidification implies an increase of the rate of erosion processes. It is typical of 80% of the region's basins. In the structure of the erosion processes, the role of storm water wash is enhanced, which is attested by positive trends of the erosion index of precipitation.

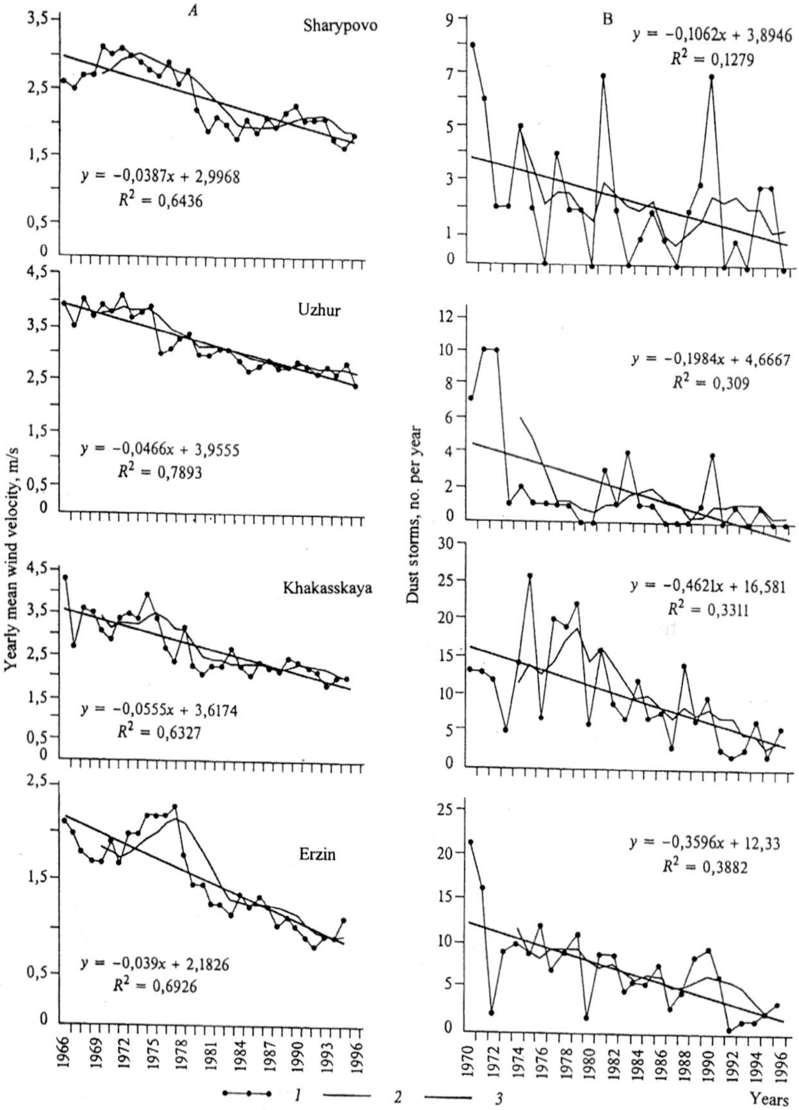

Figure 1: Long-term dynamics of wind velocity (*A*) and dust storms (*B*) in the Nazarovo (Sharypovo, Uzhur), Southern-Minusinsk (Khakasian) and Ubsunur (Erzin) depressions. *1* – yearly data; *2* – smoothed data over three-year data; *3*-trend.

The increase of the rate of erosion processes during the last 10-15 years is attested by the intensification of gullying in the Nazarovo, Southern-Minusinsk and Sydo-Yerbino depressions [13, 14]. Experimental field surveys and analyses of aerospace images and large-scale maps for different years show that gullying is proceeding at a particularly high rate (as fast as 7 meters per year) in the right-bank forest-steppe areas of the depressions. These gully systems present a significant potential threat to the preservation of the land reserves.

The second, Angara region is distinguished by the oppositely directed character of variation of the rate of the processes. In the presence of a generally descending tendency of the erosion processes, caused by the negative trend of atmospheric humidification and by the decrease in water reserves in the snow, some river basins show a marked intensification of erosion activity associated with mass timber felling. The strongest influence of concentrated felling on erosion processes is observed in felling areas not older than 2-5 years. If the felling areas occupy 1-5% of the area of the basin, the modulus of suspended sediment flow increases by 50-300%, and water turbidity increases by 100-800% [15]. Such an enhancement of erosion is characteristic for the Educhanks river basin. At the present time the volume of lumbering in some areas is increasing.

The intensification of erosion processes is also associated with the increase of the number of forest fires. The area of new burned-over places exceeds by a factor of two the area of fresh felling areas. The strongest increase in intensity of erosion processes is observed in the zone of influence of the Bratsk reservoir: in the basins of the Unga, Osa, Zalari, Oka and Belaya rivers, and of the other Angara tributaries. The mean growth rate of small erosion forms in loess-like loams is 0.2-2.0 meters per year. The last wave of global warming has contributed to the increasing recurrence of periods of a high and low water levels in the reservoir, as well as on the on-going active development of erosion and restructuring of the river banks, against the expectations of some decay of these processes.

In the third, Selenga region that encompasses the Barguzin and Selenga river basins, the descending trend of suspended sediment flow predominates. The chief reason for that is the abrupt decrease in agricultural activities on this territory during the last 20-25 years. In some regions of Buryatia, with an almost total cutting down of agriculture caused by socio-economic factors, there is a decrease of the growth rate of gullying, and the rate of sheet erosion on the slopes [13].

The fourth region (the Upper Amur basin) is characterized by the opposing tendency, i.e. an increase in intensity of erosion processes in the presence of a progressive growth of atmospheric humidification in the second half of the 20th century. The directional increase in the sediment flow in the Transbaikalia during 1970-1980 was pointed out by Bobrovitskaya [7]. The intensification of erosion processes is also characteristic for neighboring areas of the Amur region [16]. The increase in erosion activity in the Onon-Argun steppe shows up as an increasing recurrence of extreme events of sediment flow, an active transformation of small erosion landforms (water washes, rill scours, gullies). The catastrophic development of erosion processes in 1998-1999 led to a rejuvenation of the gully tops and gave rise to new forms on the steppe slopes and in the shallow valley bottoms. According to experimental field observational data at the Kharanor steppe permanent station, the volumes of slope runoff and soil wash out have increased.

Thus our identified positive trends of erosion processes are typical of the watershed basins with their high intensity (the Yenisei and Amur basins). Given the on-going climate changes, these tendencies will be still more enhanced. A most dramatic increase in sediment flow should also be expected in areas of recent economic development (gold mining, production of petroleum and gas, timber felling, etc.). In this connection, to reduce the volumes of runoff and improve the ecological situation requires measures of ecological-geomorphological safety, primarily the restoration and creation of new anti-erosion systems.

The results from calculating the trends are combined into a summary table characterizing the sensitivity of different zonal-sectoral types of morphodynamical systems to climatic fluctuations. The highest sensitivity corresponds to piedmont steppe systems, Central Asian steppe systems, and transitional forest-steppe and desert-steppe systems, where more than 75% of the parameters analyzed here are changing. Weak sensitivity to the on-going climatic changes is shown by the dry steppe systems of the Uda district, which have evolved on a thick sand-lithogenous basis (less than 25% of the parameters are changing). Moderate sensitivity is characteristic for the other types of systems.

In the analysis of the trends it was taken into consideration that each type of systems is distinguished by its "starting" (basic) functioning regimes [3] (Fig. 2, *A*). A subsequent transformation of the regimes of exogenous relief-formation takes place according to the basic level, the sensitivity of the systems to climatic fluctuations, and to the character o the climatic signal (the ratio of warmth and moisture trends (Fig. 2, *B*).

A limiting factor of changes in the functioning regimes of the systems is moisture. In areas with high growth rates of atmospheric humidification there is an increase of the energy base of relief-formation, with positive trends of exogenous processes observed here. Such a humid type of transformation of the functioning regimes of the systems is characteristic for the Koibalsky, Onon-Argunsky and Nazarovsky districts. If the climatic tendency (the increase in atmospheric humidification) continues, then for the Koibal steppe, with due regard for its basic regimes, one should anticipate the development of processes typical of forest-steppes. In the Nazarovo forest-steppe there will occur an enhancement of the role of slow mass movements of earth materials (defluxion), and landslide and avalanche-scree processes, as well as fluvial activity, i.e. the processes typical of the subtaiga territories will be enhanced.

Table

Tendencies of changes of exogenous processes caused by climatic trends

Morphoclimatic region	Climatic trends		Direction of processes							
	yearly mean air temperature, °C/yr	annual amount of precipitation, mm	basin erosion	slope storm water wash	meltwater runoff and wash	nivation	defluxion	aeolian		Permafrost degradation
								Dust storms	climatic index of deflation	
Krasnoyarsk-Kansk MC FST	0.08	0.13	+	+	o	o	-	-	+	
Nazarovo MC FST	0.08	1.06	+	+	o	o	+	-	-	
Minusinsk MC ST	0.07	1.14	+	-	-	o	o	-	-	
Koibal MC PM ST	0.05	2.50	+	+	+	+	+	+	-	
Angara-Lena SC FST	0.03	-0.95	-	o	-	-	-	+	+	+
Barguzin SC ST	0.08	0.09	-	-	o	o	o	-	-	+
Uda SC DST	0.06	0.53	-	o	o	o	o	o	o	+
Selenga-Khilok SC ST	0.05	-0.55	-	-	o	o	-	o	o	+
Onon-Argun EC ST	0.04	2.13	+	+	o	+	+	+	+	+
Kyzyl EC DSST	0.09	-0.99	-	+	+	o	-	+	+	+
Ubsunur ES DSST	0.05	-0.49		+	+	+	-	-	-	+

Note: MC FST – moderately continental forest-steppe; MC ST – moderately continental steppe; MC PS – moderately continental piedmont-steppe; SC FST– sharply continental forest-steppe; SC DST – sharply continental dry steppe; EC ST – extra-continental steppe; EC DSST – extra-continental desert-steppe. The "+" sign – increase in intensity of processes; "o" no changes; "-" – decrease in intensity.

The tendency toward a decrease in humidification is currently clearly pronounced in the Angara-Lena forest-steppe region in which the role of aeolian migration of material is enhanced, in the Selenga-Khilok steppe, as well as in the desert steppes of Tyva and the near-Olkhon area, where the transformation of the exogenous regimes follows the arid type. The high rates of warming, combined with a gradual decline in atmospheric precipitation, as well as with the anthropogenic xeromorphization and degradation of the soil and vegetation cover will contribute to an expansion of areas with the dry steppe morphogenesis (the Barguzin, Uda, Ubsunur, and Selenga-Khilok districts), while in the Kyzyl and near-Olkhon districts the present climatic trends are indicative of the likely appearance of the semi-desert type of morphogenesis.

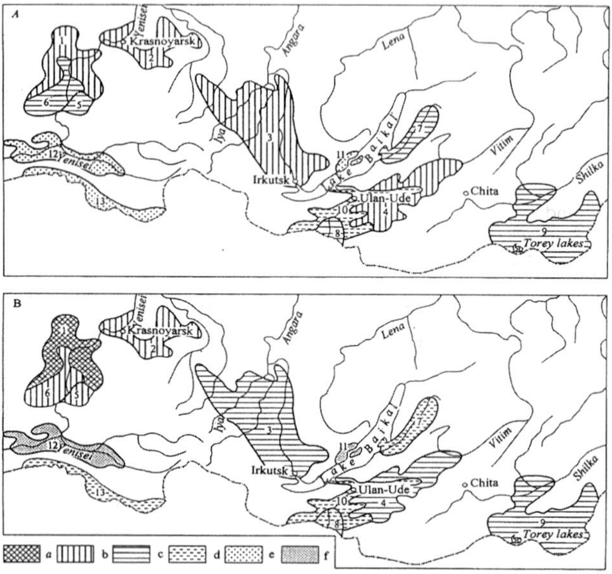

Figure 2: Basic (before 1980) functioning regimes of the morphodynamic systems of the sub-arid regions in southern Siberia (*A*), and their likely changes, with current climatic trends remaining unchanged (*B*).

Types of morphogenesis regimes: *a* – subtaiga, *b* – forest-steppe, *c* – steppe, *d* – dry steppe, *e* – desert-steppe, *f* – semi-desert; morphoclimatic regions: 1 – Nazarovo, 2 – Krasnoyarsk-Kansk, 3 – Angara-Lena, 4 – Selenga, 5 – Koibalsky, 6

– Khakasia, 7 – Barguzin, 8 – Selenga-Khilok, 9 – Onon-Argun, 10 – Udinsky, 11 – near-Olkhon, 12 – Kyzyl, 13 – Ubsunur.

Thus the erosion, aeolian and cryogenic processes of the sub-arid landscapes are distinguished by high sensitivity and rapid response to the on-going warming. Overall, under the influence of current regional changes of climate within the zone of island steppes and forest-steppes in southern Siberia, there is an increase in diversity and contrast of the regimes of exogenous relief-formation.

This work was done with financial support under the Interdisciplinary Integration Project SB RAS (138).

REFERENCES

1. Timofeyev D.A. A new stage in the activity of the Geomorphological Commission of the International Geographic Union. Geomorfologiya, 2001, No. 3.
2. Simonov Yu.G. Geomorphology. Methodology of Basic Research. SPb: Piter, 2005.
3. Bazhenova O.I. and Martianova .N. Current morphoclimatic regimes of the steppes and forest-steppes of Siberia. Geography and natural resources, 2002, No. 3.
4. Bazhenova O.I. Large-scale mapping of the dynamics of current exogenous processes. Ibid., 1993, No. 1.
5. Mitchell J.K. and Bubenzer G.D. Chapter 2: Soil Loss Estimation. In M.J.Kirkby and R.P.C.Morgan (eds): Soil Erosion, John Wiley & Sons, Chichester, UK, 1980.
6. Borovitskaya N.N. An investigation and calculation of soil wash out from the slopes. A Collection of papers on Hydrology. Leningrad, 1977, No. 11.
7. Borovitskaya N.N. Water Erosion on the Slopes, and River Sediment Flow: Abstract of Dissertation (Dr.Sc. (Geogr.), SPb, 1995.
8. Wilson S.J. and Cooke R.U. Wind erosion. In M.J. Kirkby and R.P.C. Morgan (ed.) Soil Erosion. John Wiley & Sons, Chichester, UK, 1980.
9. Grechishchev S.E. Forecasting the permafrost thawing and distribution, and of changes in in cryogenic soil fracturing on the territory of Russia at the period of climate warming. Kriosfera Zemli, 1997, v.1, No.1.
10. Devdariani A.S. Measuring the Movements of the Terrestrial Surface. Moscow: Nauka, 1964.
11. Gusarov A.V. Trends of Erosion of Sediment Flow on the Earth in the Second Half of the 20ttth Century: Abstract of Sisertation (Cand.Sc. (Geogr.). Kazan, 2005.
12. Bazhenova O.I. and Martianova G.N. An assessment of changes in geocryological conditions of the sub-arid regions of Siberia at the period of current climate warming. Geography and natural resources, 2003, No. 4.
13. Bazhenova O.I., Lyubtsova E.M., Ryzhov Yu.V., and Makarov S.A. Spatio-Temporal Analysis of the Dynamics of Erosion Processes in the South of East Siberia. Novosibirsk: Nauka, 1997.
14. Kozhukhovsky A.V. Gullying in the Sydo-Yerbinskaya Depression: Abstract of Dissertation (Cand.Sc. (Geogr.). Krasnoyarsk, 2004.
15. Burenina T.A. and Onuchin A.A. The runoff dynamics in the watershed areas of the Angara region in connection with timber felling and forest fires. Geography and natural resources, 1999, No. 2.
16. Makhinov A.N. and Makhinova A.F. Influence of global climate change on current gullying in the south of the Far East. New and Traditional Ideas in Geomorphology: proc. 5th Shchukin Lectures. Moscow: Moscow University Publisher, 2005.

CHAPTER

4

Landscape-Hydrological Analysis of A Territory

A.N. ANTIPOV

Not only does man's influence on the natural environment change natural landscapes, but it adversely affects a runoff formation within their boundaries. Research experience in this field shows that anthropogenic transformations of natural complexes lead as well as to primarily qualitative depletion of water resources and also to unfavorable redistribution of river stow over: a decrease in drought river discharge and a considerable enhancement of hazardous hydrological phenomena. The Russian Federation has unique reserves of pure fresh water whose significance may well compare with energy resources. Loss of this resource could become a strategic challenge for the country in the foreseeable.

Resolving the water-ecological problems is impossible without a sophisticated understanding of hydrological processes of territories. Over integration and too formal views of hydrological processes and phenomena is one of the main indicators of poor hydrological studies. Therefore, an urgent need for a further differentiation of the concepts regarding hydrological organization of territories is obvious. Developing of a new point of view concerning the study of hydrological processes and phenomena is important to solve the aforementioned problems. Firstly as regards the estimation of the influence of varied and interrelated natural factors and natural structures on hydrological processes, i.e. an understanding the hydrological territory organization of a different spatial dimension with different physio-geographical conditions and a different type of economic use. Secondly, as a description of hydrological properties of natural systems as such, including environment-forming functions of water bodies.

THEORY AND METHODS OF LANDSCAPE-HYDROLOGICAL ANALYSIS OF A TERRITORY

The elaboration of concepts of hydrological functions of landscapes and their separate elements centers around geographical sciences. A historical overview of the problem is given in [1].

Theoretical ideas of landscape-hydrological territory organization are based on V.B. Sochava's system paradigm. According to this paradigm, the environment is represented as an hierarchy of different-type and different-scale and different-sized geosystems. Applied to hydrological processes and phenomena, these ideas are transformed relative to the water component of the geosystem which V.B. Sochava treated a critical component that predetermines the functioning of the system as a whole.

Taking into account the obvious interrelation and interdependence of hydrological processes and phenomena and natural factors, a definition of one of the types of partial geosystems – landscape and hydrological (LGS) is offered. By them is meant the portion of the terrestrial s surface where the interactions of hydrological processes and natural structures are conditioned and localized by specific regularities. Each spatial level and type of LGS has its own type of interactions.

The climate predetermination of river discharge from territories of a global dimension, the significance of the swamp micro-landscape or the degree of drainage – for a typological dimension can be taken as a representative example.

Table 1 gives an idea of the dimensions and spatial combination of geosystems of water basins and LGS. It follows from the table that between the boundaries of these geosystems there is no clear spatial correspondence. Thus on the boundaries of small river basins it is possible to single out a spatially-conjugate set of stows, while a basin itself can be ascribed to a definite locality or area. Within the boundaries of the river basins it is possible to identifier quite diverse watershed complexes formation zones of river discharge, slopes – portions of water transit into valley systems etc. These are referred to territories of accumulation and drainage. Furthermore, according to the type of basin (mountain, plain, taiga, steppe, etc) peculiarities of the processes will be quite varied. There is no one-to-one correspondence between LGS and landscape differentiation elements, though the foundation of LGS identification is the concept of geosystem predetermination of hydrological processes.

Table 1

Hierarchy of different natural systems

Spatial dimension	Basin area in thous. km^2	Order of river basin	Geosystem	LGS
Planetary < 1:5 000 000	> 300	VII – IX	Zone	Landscape-hydrological subcontinent
			Subcontinent	
	10 – 300	VI – VII	Region	
			Nature zone	Landscape-hydrological zone
Regional 1:200 000 – 1:5 000 000			Subzone	
	2 – 10	V – VII	Province	L.-g. province
	0,4 – 10	III – V	Landscape	
			District	L.-g. district
Topological 1:10 000 – 1:200 000				L.-g. subdistrict
	< 0,4	I – III	Locality	L.-g. locality
			Stow	L.-g. site
			Facies	L.-g. fundamental site (fundamental flow forming complex)

The definition of the main units of the hierarchy is provided, based on research experience under different natural conditions and at different spatial levels. For instance, a landscape-hydrological province is a single system of the size of a region, which reflects the modification to the zonal type of landscape-hydrological patterns, caused by the spatial-temporal overlap of climatic conditions (heat and moisture availability), and large orographic landform elements or morphostructural entities with a definite attribute. A combination of the interacting processes and natural structures determines the specific composition of hydrological objects: the specific character of the drainage network structure and the forms of river basins, the genetic and morphological characteristics of lake formations, the predominance of the type of swamp micro-landscape, and the basic forms of occurrence and run-off of gravity groundwater.

The territory of a province has a homogeneous proportion of water balance elements, with a single direction of hydrological processes predominating.

According to the spatial level, LGP corresponds to a river basin of order III-IV.

The identification of an elementary homogeneous area, the biogeocenosis, is obvious. This elementary unit corresponds in some cases to the elementary flow-forming complex (SFC). The union of biogeocenoses gives an idea of the following hierarchic unit: an elementary slope or a simple stow, within the confines of which there are no natural boundaries, primarily orographic.

The above general theoretic concepts are based on the general framework of landscape-hydrological analysis. The chief goal of the analysis is to determine its object. In a spectrum of hydrological problems it is possible to single out quite different classes of them. A most common purpose implies determining the hydrological organization of the territory or its elements.

Estimating the hydrological organization often forms the basis for the more specific and practical tasks. For instance, a typical task is to map the hydrological characteristics. A key component in the purpose-oriented planning effort is the scale of LG analysis. It is determined by the size of the area of the territory analyzed, and by the required mapping detail of the processes and phenomena.

An important element of the LG analysis procedure is mapping the ratio of hydrological to natural systems. The procedure of "feeding the basin" with geographical information is considered the most complicated for hydrologists because orographic and complex natural boundaries do not coincide. It is not advisable to superimpose the geographical mosaic on orography; instead. It is necessary to determine the watershed boundaries on complex and specialized maps.

The switch from the concepts about the information of hydrological processes and phenomena within river basins to their definition within the boundaries of natural systems relies different methods. On the whole they may be divided into empirical and indicational. The former approach is based on experimental data [3]. In this case representative sites are provided with appropriate measurements of water balance elements. Next, relations between the resulting values and natural factors are established, and these relations are used for extrapolation in the natural systems under study.

The latter approach is largely based on the methods of hydrological indication and typology. As a rule, the main argument of calculations involves the characteristics of river discharge in the outlet of a river basin, and the result involves the characteristics of flow formation in single elements of a complex or specialized differentiation of a basin [1].

The final level of LG analysis involves identification of spatial units of hydrological organization at a level of particular and complete LGS taking into account methods of typification and zoning.

LANDSCAPE-HYDROLOGICAL ORGANIZATION OF RIVER BASINS

Basin systems refer to basic units of flow formation. Traditionally a river basin is treated as a uniform surface, characterized by a set of different, yet common (to its area)| characteristics. An alternative line of inquiry includes research into the interrelation of hydrological processes and natural structures within the boundaries of river basins as well beyond them. In this case the internal structure of river basins is revealed, i.e. a basin is viewed as a spatially ordered system of natural complexes, having different hydrological functions, which involves the procedure of hydrological zoning of basins.

a – loose deposits and bedrock (according to I.P. Rabshe): 1 – rubble deposits on rubble basement rocks, 2 – boulder-shingle deposits on rock, 3 – rubble deposits on rock, 4 – boulder-shingle deposits on rubble bedrock, 5 – rubble deposits on transitional rock with the inclusion of clay and rubble material, 6 – rubble deposits on the old weathering crust, 7 – clay deposits on transitional basement rock, 8 – clay deposits on the old weathering crust, 9 – sand deposits on transitional basement rock;

b – transpiration discharge by plant communities, mm: 1–50–100, 2–101–150, 3–151–200, 2–201–250, 5–251–300, 6–301–350, 7–351–400, 8–401–450;

c – hydrological regime of biogeocenosis: 1 – types of boundaries, 2 – subtypes of boundaries, 3 – type index, 4 – subtype index;

d – territorial distribution of water abundance (l/c km^2) from territory of biogeocenosis (high water in 1977): 1–25, 2–50, 3–100, 4–> 200.

Landscape-hydrological analysis of different-scale river basins can be reduced to searching and target estimation of landscape characteristics, which indicate a hydrological cycle, or can be conducted during specialized experimental water balance investigations in elementary natural complexes of basins concurrently with the determination of the main factors of formation of hydrological processes. The analysis is normally based on thematic specialized or complex landscape maps that reflect and indicate the hydrological organization of the territory.

A detailed description of the experimental basin of the Balakhta river (area 13.6 km^2) is given in a number of publications [3]. The basin lies in the contact zone of mountain topography of the Western Sayan mountains and plain topography of the Southern-Minusinsk depression. Its territory covers the first ridges of the Western Sayan and is characterized by a significant height differentiation, ranging from 1200 m in the basin area, to 550 m at the outlet.

Comparative analysis of hydrological and other thematic maps of the basin of the Balakhta river and its tributaries enables a typification of the hydrological regime at a level of biogeocenoses and basin slopes (Fig. 1). Types of biogeocenoses regimes were singled out with due regard for the character of gravity water runoff, and subtypes are based on amount of moisture losses for evaporation, mainly for transpiration.

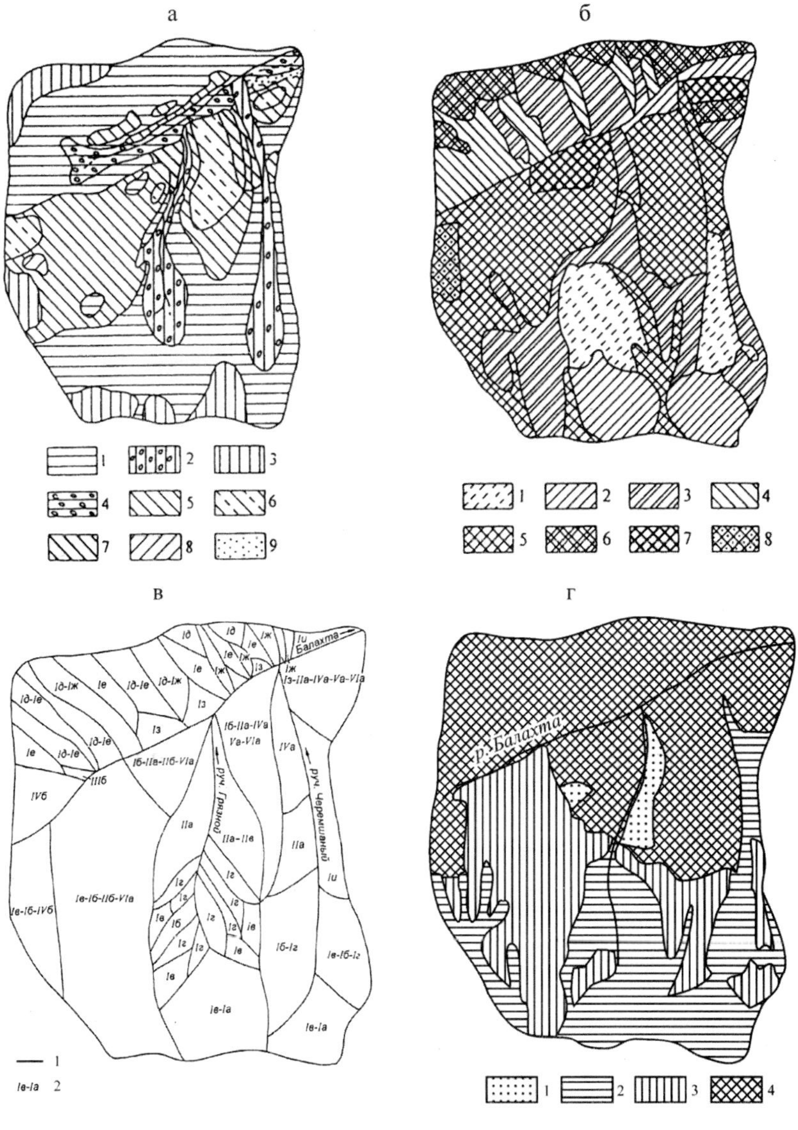

Figure 1: Schematic maps of the Balakhta river basin.

The combination of hydrological regime types of biogeocenoses within the boundaries of slopes is regulated according to a differentiation of loose sediments and underlying rock, i.e. it is largely determined by the slope geometry. Therefore the most complicated structure of the regime (for example, conjugation of types Ic – Ib – Iib – Va) is formed on downweared convex slopes of the Balakhta river basin. A simpler structure was observed for weakly convex slopes of the Cheremshanny and Gryaznoy stream basin (union of types Ic – Ia).

The structure of the hydrological slope regime determines regularities of runoff formation (Table 2). Maximum water content in the low-water period is observed at weakly convex slopes, and minimum water content is observed at concave slopes from which there is no the gravity water runoff or it is only minor. In moisture-abundant periods, in spring or during steady rainfall in summer, the character of runoff formation for such slopes is changed. On the slopes where the structure of the hydrological regime is represented as Ic – Id or Ib – Id union, the water content increases 3–5 times. On the slopes with more complicated union types of biogeocenose regime (Ic – Iia – Iva – Via or Ic – Ib – Iid – Via), and the gravity water runoff increases by 50–100 times.

The dynamics of river basin water content also totally depends on the type of hydrological regime. Two tributaries of the Balakhta river are characterized by a change in water content for slopes where hydrological regime I is of crucial importance or II is of less importance and water content of the Balakhta river basin without tributaries is determined by a more complex union of types (for example, II – IV – VI). As a consequence, the runoff of the Balakhta river forms at the outlet while the low-water period in basins of its tributaries, the Cheremshanny and Gryaznoy streams (see Fig. 1) and during the high-water period and summer floods the role of runoff formation increases dramatically in the rest of the basin. There is long-term average water content that has maximum values at the Gryaznoy stream (20 l/c-km²) and minimum values at the Balakhta river without tributaries (8 l/c-km²), in accordance with the structure of the hydrological regime of the basins.

Table 2

Slopes water content of the Balakhta river basin, l/c km²

Slope index	Area km²	Hydrological phase		
		Low-water period	Summer high water	
		20VI 1977	23VI 1976	12VII 1977
Ic-Ia, Ic	0.56	48.2	21.4	143.0
Ib-Id	0.51	25.5	29.4	98.0
IIa-IIc-IIa	1.34	7.5	1.0	275.0
Ii, Iva	1.69	4.7	1.2	136.1
Ic-Ib-IIb-VIa Ic-Ib-IVb	2.29	1.4	1.1	148.8
Ic-IIa-IVa-VIa Ia-IIa-IVa-Va-VIa	2.06	12.0	7.2	310.0

Tree felling took place in the area of more than 20 % of the total area of the basin during the period 1977–2000. As a result, the landscape and hydrological structure of the river basin was totally altered in the areas affected by fellings, where the runoff increased 15 times during the flood period. At the same time, water losses from slopes, which were not affected by fellings did not change. During the recession of flood and at the passage to the low-water period the situation was opposite. The runoff from felling near watershed areas decreased drastically and reached almost zero-point values by the low-level water period. According to the measured discharge of water and subsequent calculations, the dry-weather flow forms mainly due to the portions of the slope not affected by felling and covered with dark coniferous plant communities.

WATER PROTECTION ZONING OF THE SHORE AREAS OF LAKE BAIKAL

Not only the most severe water-ecological situations but also various conflicts between land and water users occur on territories of water protection areas. The most landscape-hydrological systems organized in a most complex way are presented here as concisely as possible.

Legalization of water protection zones in Russia is largely formalized and is reduced to the relations between the length (water course) or the area (lake and water reservoir) of water bodies and the width of water protection areas, which enable a preliminary minimal estimation of water protection areas for a given water body. A subsequent procedure of water protection zoning should be based, on the one hand, refinement of the outer boundary, and, on the other, on its internal differentiation according to the system of natural-ecological attributes. The target function for the zoning of the Baikal shore areas is preservation of shore territories of Lake Baikal from which pollutants enter into Lake Baikal in the range of their natural variations, and preservation of self-purifying and runoff forming capacity of rivers and landscapes on their basins.

The concept of water protection zones is largely connected with the realization of a runoff (pollutants included) over the surface, i.e. it is necessary to determine the length of the route where total infiltration of water entering the ecosystem (linear size of water protection zoning) will take place or the area from which the sheet flow is possible (geometry of water protection zoning). Knowing the structure of landscapes, adjoining a water body, and the characteristics of its elements, it is necessary to estimate the mass and energy exchange and the flows to the base of drainage by the computational or expert method.

A definite sequence of steps is suggested for designing water protection zones. The target function of water protection zoning is formulated at early stages, based on analyzing the water-ecological or natural-geographical problems as well as legal provisions. The second stage is informational and analytical, which includes research into single environmental units in the context of their influence on the hydrological organization and water-ecological situation. In conformity with the zoning of the Baikal shores and taking into account the target function as defined, the following environmental units: the soil and vegetation complex, the geological-geomorphological and hydrological system, and landscape forms of territory differentiation.

Thus, in designing water protection zones, the geomorphological factor is taken into account for singling out of the landscape high-altitude zones, and the positions and exposures (solar and gravitational) of landscape elements, and largely corresponds with typological characteristics of forests and genetic soil types. A number of soil characteristics is to estimate the potential of soils as regards the water regulation capacity. They include granulometric composition, total porosity, hydrological capacity, water permeability, water regime type etc. Geochemical criteria of the water protection function are determined by the potential sorption capacity of soils. Such soil characteristics as acid-alkaline and redox conditions, the cation absorption capacity, and amount of exchange bases in horizon A, presence of carbonates and geochemical barriers, permafrost. A possibility for direct losses of river flow in the shoreline area, and the degree of protection of ground-water in connection with conditions of their formation have been investigated.

A critical stage involves the conversation of specialized assessment of natural environmental units to parameters of hydrological functions, i.e. carrying out a landscape-geographical territory analysis. Results of such analyses enable one to proceed water protection zoning as such. Furthermore it is important to focus on the solution of water protected tasks in watershed areas, rather than in rivers and water bodies, i.e. the prevent, rather than on liquidation of pollution aftereffects.

The research done enabled an assessment of the water protection potential of different landscape elements and its variation depending on the character of technogenic influence in perspective. Integral assessment of the water protection potential is also a water-ecological scale used to identify typological elements of a water protection zone. Integration of typological elements permitted the determination of the boundary of the water protected zone of Lake Baikal and the internal (Fig. 2) as a basis for various water protection measures and water-ecological regulations of economic activity.

The situation for water protection zoning of residential centers of the shoreline territory of Lake Baikal is different [4]. In spite of their obvious significance as a critical source of influence on water bodies, minimization of water-ecological effects is mainly based on technological cleaning schemes for runoffs of a different origin. The self-cleaning potential of a territory is overlooked in the planning of water-economic measures.

The setting-up of the water protection framework of territories of residential centers may be regarded as a general line of inquiry for solving water-ecological problems of urban territories. This implies a system of territorial limitations of human activity, ensuring a standard quality level of surface and ground-water, flowing into Baikal (or some other water body) as well as maintenance of the optimal drainage regime and minimization of erosion, abrasive, debris, icing and other hazardous hydrological processes (Fig. 3). The town of Baikalsk, situated within the shoreline area of Lake Baikal, is an example of water protection zoning. Thus the suggested landscape-hydrological concept could aid in solving water-ecological problems, based on basin management and focusing on preventive measures.

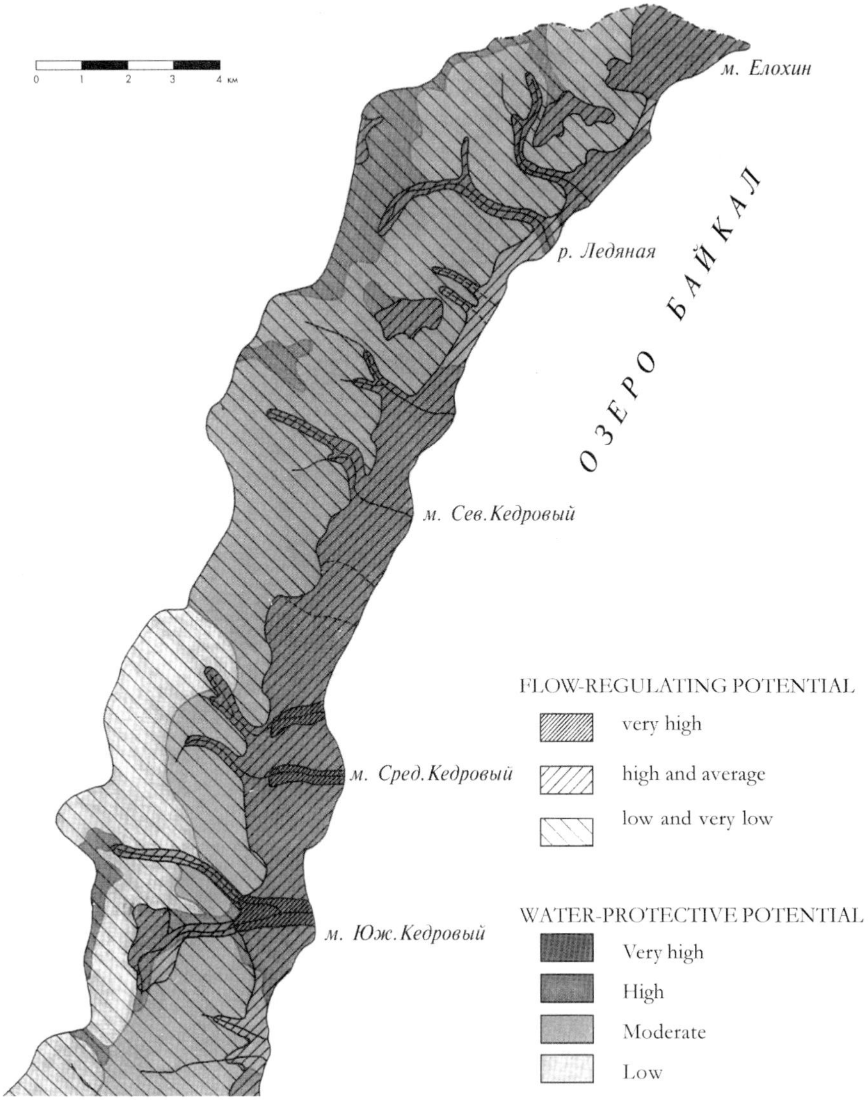

Figure 2: Landscapes flow-regulating functions of a section of Lake Baikal water protection zone.

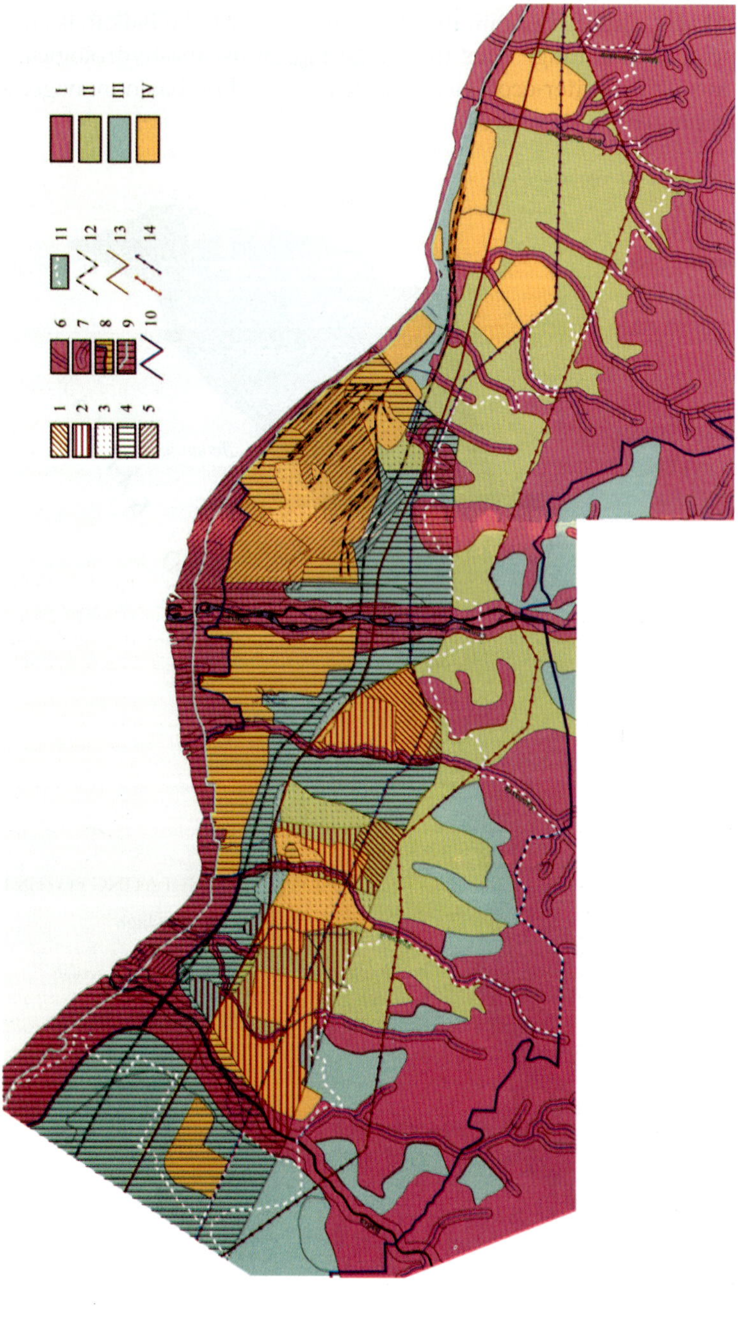

Figure 3: Schematic of water protection zoning for the territory of the town of Baikalsk.

Water-protective potential of underground water		Characteristic of the degree of protection of underground water			
1	Very high		The underground water recharge rate is very low at high or very high seepage flow rate		
2	High		The underground water discharge rate is very low and low at high seepage flow rate		
3	Moderate		The underground water discharge rate is low and moderate at high and moderate seepage flow rate		
4	Low		The underground water discharge rate is moderate and high at moderate and low seepage flow rate		
5	Very low		The underground water discharge rate is very high and high at very low and low seepage flow rate		
6	Permanent streams and water protection zones	7	Temporary streams and water protection zones		
8	Water protection zone boundary	9	Coastal-protective strip of Lake Baikal		
10	Central ecological zone boundary as per the law "On the protection of Lake Baikal" (draft law)				
11	Boundary of the envisaged natural zone				
12	Railroad lines	13	Automobile roads of federal significance	14	Electric power

Zones of territory development		Brief characteristic of the zones	Principles of nature management
I	Preservation of the existing status	Runoff forming and water protection natural-territorial complexes represented by dark-coniferous, steep-slope, humid, near-fluvial and swampy complexes in the Baikal's shore areas	Preservation of the existing landscape structure and its integrity; abandoning activity causing topsoil and overland flow disturbances
II	Intensive development	Small-leaved, sparse-forest natural-territorial complexes and wastelands on plain and inclined surfaces not drained by streams with high and moderate percolation and sorption properties of soils	Impacts on landscape-hydrological complexes should secure possibilities for the recovery of water balance in the range of its natural variations
III	Extensive development	Small-leaved, sparse-forest natural-territorial complexes on surfaces with moderately steep slopes and high and moderate water permeability of soils	Impacts on landscape-hydrological complexes should be regulated with due regard for antierosion measures; provision for collection and recovery of pollutants with a total exclusion of their ingress onto the soil surface; conduct of landscape planning with due regard for the water factor on the territories of planned development

IV	Preservation of the current use with elements of improvement	Territories of urban and industrial-economic development with a high level of disturbance of natural-territorial complexes and pollution, the source of contaminant ingress into streams and underground water	Recovery of qualitative and quantitative characteristics of surface and underground water, approaching background ones; Nature conservation optimization related to the avenues for development of housing and communal services, and the industrial economy; Observance of the water protection zoning criteria when allocating gardening cooperatives and recreation zones; Prohibition of any allocation of pollution sources unless appropriate steps to block off their routes of ingress into water bodies and aquifers

References

1. Antipov A.N. Fedorov B.N. Landscape-Hydrological Territory Organization. Published by SB RAS, 2000, 254 p.
2. Sochava V.B. Research on Geosystems. Novosibirsk: Nauka, 1975, 39 p.
3. Antipov A.N., Korytny L.M. Geographical Aspects of Hydrological Research. Novosibirsk: Nauka, 1981, 177 p.
4. Antipov A.N., Gagarinova O.V. Water Protected Zoning of Residential Centers. Geography and Natural Resources. 2002, No. 4, pp. 30–36.

CHAPTER

5

Landscape and Landscape - Geochemical Research

V.A. Snytko and Yu.M. Semenov

The advancement of the theory of science elicited in the 1960s the emergence of new ideas in physiography. They were tied to system concepts that gained acceptance in many countries in connection with the formulation of L. von Bertalanfi's general theory of system analysis. The quest of a new concept was dictated by the need to reconcile route and stationary investigations and to organize them at a new level complying with the modern trends in geography. To do this required arranging according to a system a number of theoretical postulates that took shape in landscape science and invoking additional ones from related disciplines, primarily from ecology.

The emergence of the science of geosystems goes back to 1963 when V.B. Sochava [1] coined the term "geosystem" and introduced it into scientific usage. Subsequently the author defined this notion as follows: "...the geosystem (irrespective of its dimension) is a whole entity consisting of interrelated components of Nature that obey regularities operating in the geographical envelope or landscape sphere" [2, p. 4], or "...using the language of cybernetics it can be said that it is a special class of controlling systems; the terrestrial space of all dimensions where the individual components of Nature are in system linkage with one another and, as a certain integrity, interact with the cosmical sphere and with human society" [3, p. 292].

Initially, the new branch of science was treated as structural-dynamic landscape science [4]. The fundamental ideas pertaining to the science of geosystems were advanced by V.B. Sochava at the 5th Congress of the USSR Geographic Society in 1970 in his paper entitled "Geography and ecology" [5], and at the 6th Congress in 1975 his paper "Science of geosystems" was presented [6]. The geosystem paradigm came to gain wide acceptance when V.B. Sochava's papers were published in the journals of "Izvestiya VGO" [7] and "Petermanns Geographische Mitteilungen" [8, 9].

In a final form, the most important postulates of this science were set forth by the author in a monograph entitled "An Introduction to the Science of Geosystems" [3]. Geosystems represent a special class of open, hierarchically organized, dynamic systems. The hierarchical character of structure is their critical property because, thanks to it, both an elementary area of the terrestrial surface and the planetary geosystem constitute a dynamic entity with a special geographical organization inherent in them. Organization of geosystems includes their differentiation, integration, development, and sustaining functioning; therefore, the science of geosystems involves, as ingredients, theoretical postulates that substantiate the regularities of their classification, mapping, dynamics and evolution.

Among the basic objectives achievable on the basis of the science of geosystems, are: modeling of geosystems with consideration for their spontaneous and anthropogenic dynamics and natural regimes corresponding to them; the quest of rational techniques for quantitative assessment of geosystems and landscape-forming processes; system analysis of spatial links in the geographical envelope at the planetary and regional levels; understanding of the spatial-and-temporal regularities, and analysis of the states of geosystems; creation of graphic models of geosystems, primarily maps of the habitat in connection with problems related to its protection and optimization; study of the influence of socio-economic factors on the natural environment, and prediction of future geosystems; development of the foundations for constructive landscape science in order to generate "man-environment c-creation geosystems", one of the forms of target-oriented anthropogenic transformation of geosystems in accordance with evolution trends inherent in them [3].

A very important consideration for understanding the essence of geosystems is the notion of the invariant, a totality of properties of the geosystem that remain unchanged as it is transformed under the influence of external factors. Cyclic changes of natural properties and regimes that occur for a definite time and do not leave the confines of the invariant, are called the dynamics of geosystems, while radical alterations that lead to a change of the invariant, are termed the evolution. Both dynamics and evolution represent links of the same process of development of geosystems – dynamic (quantitative) changes that accumulate in natural entities would lead to a qualitative jump, the evolution. The interaction of the components in geosystems is sustained by flows of matter and energy, and the entire totality of the processes of exchange and transformation of matter and energy is referred to as the functioning of geosystems.

The first research ground where the ideas forming the basis of the science of geosystems were tested was the Kharanorsky steppe physiographical permanent station in the South-Eastern Transbaikalia the principle objective of the work done under the guidance of V.B. Sochava was to pursue a comprehensive physiographical investigation of the steppe geosystems including the study of rhythms of natural phenomena, substantiate a classification of natural entities, and to identify structural-dynamic properties inherent in them [10].

Subsequently, the scope of stationary investigations was expanded – they incorporated the substantiation for activities aimed at transformation of the natural environment, the development of principles and methods of physiographical forecasting, and identification of potential lands of various kinds of economic utilization. To solve the problems of approaching an integral quantitative assessment of natural regimes that would open the way to settle a number of theoretical issues related to regulation and transformation of the natural environment, and to make physiographical predictions required not only estimating the parameters of flows of energy, biomass and inorganic matter but also studying the organization of subdivisions of the geographical environment. The stationary investigations were carried out, based on using the method of comprehensive ordination, the essence of which is a syntope and synchronous study of the natural regimes in transect-research areas in order to identify natural combinations as systems changeable in space [11]. By studying separate natural regimes and integrating them and identifying cycles of chemical elements in contiguous geosystems, it was possible to establish the major dependencies inside steppe geosystems through correlation and regression-variance analyses. They formed the basis for the creation of graphic and mathematical models of elementary geosystems, and identifying critical components of the steppe landscape: effective radiation, circulating moisture, and productivity of grass [10].

A large body of work done on investigating the structure and regimes of geosystems at Siberian geographical permanent stations made it possible to determine and unravel spatial differences of the functioning of geosystems with due regards for the temporal dynamics of biotic and abiotic factors in the steppes of the South-Eastern Transbaikalia [10, 12] and Yuzhno-Minusinskaya depression [13], in the taiga of the piedmonts of the Western Sayan [14], the Krasnoyarsk Angara region [15, 16], the Southern Irtysh region [17-19], and the Kondo-Sosvinskoye Ob region [20].

A topological study of the landscape of the Upper-Chara depression [21] made it possible to carry out its detailed physiographical facies analysis with consideration for the chorological links between facies within a district and for the character of the processes which determine the formation of their factorial-dynamic series and various combinations.

Within the context of elaboration on the science of geosystems, A.A. Krauklis, on the basis of investigations from the Priangarsky taiga station, created the framework for a new working area of comprehensive physiography, experimental landscape science that is at the junction of landscape mapping, regionalization and special-purpose study of the environment using methods from exact sciences [22]. The author suggested the concept of the geosystem as a set of functionally linked variable states of the natural complex, developed the landscape model in the form of a system of factorial-dynamic series of facies, and examined the prospects and avenues for practical implementation of experimental landscape research for purposes of development of the taiga.

This monograph has much in common with a book by V.V. Ryumin [23] which, on the basis of materials from the study of the landscape structure of the surroundings of the Lensky taiga station and the Novonikolayevsky steppe station, addressed the issues related to the dynamics and evolution of geosystems and their stability, and made an analysis of the natural-resources potential of the territory.

A significant contribution to the geographers' views regarding the ordering of the structure of the landscape sphere was the concept of a three-series classification of geosystems [3] that includes the typological (geomers), chorological (geochores) and dynamic (variable states and epigeomers) series. The tiniest cells of the geographical envelope are represented by elementary homogeneous (uniform) areas of the natural environment: elementary geomers or biogeocenoses. By the principle of similitude, they are combined into facies, and further, with consideration for generalization, into groups of facies, classes of facies, geoms, groups of geoms, classes of geoms, and larger taxons thus constituting a classification class of geomers. The least (as regards the number of components) territorial combination of elementary geomers represents an elementary geochore, or an elementary heterogeneous area. According to territorial contiguity and commonality of the natural properties, elementary geochores are combined into microgeochores (stows), mesogeochores (localities), topogeochores (areas), macrogeochores (districts, landscapes), provinces, etc., forming in the aggregate a classification series of geochores. These series are both interrelated, and the highest taxon of either series is the landscape envelope in general. All geosystems are subdivided into three orders of dimensions: topological, regional, and planetary. The landscape in the hierarchy of geochores occupies a nodal place – it is simultaneously a unit of both topological and regional dimension.

The graphic, or as V.B. Sochava said at one time, full-scale, model of the geosystem is represented by the landscape map summarizing results of field work. Mapping of geosystem is, in its essence, opposite to mapping of separate components of the natural environment where the primary objective is to reflect local peculiarities arising as the result of a spatial differentiation of large component formations. In the case of mapping of geosystems, however, first and foremost it is necessary to take into account the back side of the overall physiographical process – the integration of separate components into local geosystems [24].

A landmark in landscape mapping was the publication of the map "Landscapes of the South of East Siberia" [25] compiled by V.S. Mikheev and V.A. Ryashin under the guidance of V.B. Sochava; in its legend the entire totality of mapped geoms and other geomers is subordinate to those geosystems of planetary dimension to which they pertain. The map also reflects large regional taxons to which mapped groups of geoms are subordinate, as well as dynamic categories of groups of facies.

A significant place in the science of geosystems is occupied by the problem of geographical forecasting that was treated by V.B. Sochava [26] as a very important direction of modern geography and was defined by him as the development of ideas about natural geographical systems of the future. It is aimed at prediction of changes of geosystems in the process of spontaneous development and as a consequence of man's activity, and is intimately linked to standardization of nature management.

It is A.G. Isachenko's opinion [27] that landscape forecasting is a further development of V.B. Sochava's theory of dynamics and evolution of geosystems. On the other hand, the forecasting problem is directly related to issues of information support and modeling. Therefore, the Siberian geographers' forecasting developments aimed at optimization of nature management in particular regions take into account primarily the natural and anthropogenic landscape trends that were identified as part of station-based investigations, archival and published materials, and experimental data on the determination of standard loads on geosystems.

A book by V.S. Mikheev [28], dealing with the implementation of landscape research results in working out comprehensive problems related to development of Siberia's regions, examines a large system of support as the region of combination of the theory and practice of physiographical investigation in the development of territorial systems for scientific preparation of production operations of man in his habitat.

The beginning of studies on the influence of socio-economic factors on the natural environment in the light of the science of geosystems was marked by the appearance of a collection entitled "Topological Aspects of the Science of Geosystems" [29]. Subsequently, Siberian geographers addressed themselves repeatedly to the synthesis of economic-geographical and physiographical data in forecasting geosystems of the future in areas of intense development.

On the basis of integral usage of the science of geosystems [3] and the basic postulates of landscape geochemistry [30–33], staff members of the Institute of Geography SB RAS identified a large number of aspects of differentiation, integration and development of natural entities through the study of differentiation and migration of their substance component.

Landscape-geochemical research was directed at the study of metabolism processes in geosystems [33, 34]. The emphasis was on studies at comprehensive physiographical stations [35]. The objects of observation were representative (for landscape) elementary subunits, facies, that form conjugate series in accordance with the direction and extent of flows of material. In practice, account was taken of the main forms of substance state: biogenic, solid, liquid, and gaseous.

It is possible to gain insight into each of the aforementioned substance facies of geosystems in the course of special-purpose experiments. because of the complexity of the study of the solid phase, the observations of this phase of substance had to be organized mainly in a static form. Taking into account the bulk mass of the components of geosystems provided a clear understanding of the stock of substance in solid phase, which subsequently made it possible to make various comparisons and more thoroughly represent the geosystem.

Especially noteworthy is the liquid state of matter, chiefly the transformation of moisture in soils and plants, and migration of materials dissolved in the components of geosystems. In investigating the integral role of the alkali-acid conditions, attention was paid to pH dynamics.

The organic component of geosystems as the basis for all vital processes that governs the course of most natural phenomena, has a special role. Therefore, special-purpose experiments were conducted to take into account living substance and migration of chemical elements in it. It was important to detect the dynamics of substance, its seasonal (intra-annual) variations, as well as revealing year-to-year trends of the latter.

Intra-landscape transformation patterns of the gaseous phase of matter were considered in the example of soil air carbon dioxide as an indicator of the course of long-term natural processes. The study revealed its variations within definite geosystems, and transformations in their conjugate series.

A study of the vertical differentiation of substance of conjugate facies made it possible to determine the dynamic trends of landscape caused by the distribution of substance in connection with surface nonuniformity. The study demonstrated the intra-secular differentiation of substance in geographical facies on the basis of studying the series of macro- and microelements.

Geochemical investigations into landscape form part of physiographical research to ascertain the structural-functional organization of geosystems. Appropriate routine observations of the variability of substance provide not only spatial but also temporal regulararities. Results of seasonal and year-to-year observations of the particular indices at the transect-research site (conjugate series of facies) were plotted as a graph that was regarded as a spatial-temporal model of the facies.

Investigations of radial migration of material with water flows at stations used the lysemetric, chromatographic and micromorphological methods.

An outline is given of the experience gained from investigating the dynamics of taiga geosystems (by considering an example of the West-Siberian region) using informative landscape-geochemical indices. Calculating the relative parameters of the major natural processes (biogenic and water migration of chemical elements and their differentiation in the soil thickness) is treated as the quest of the form of expression of the landscape dynamics [36].

An analysis was made of the entry of technogenic material into landscapes, and of the processes of its transformation and distribution in the components of environment. The functioning and transformation of landscapes evolving in technogenic conditions were studied [37].

Station-based landscape-geochemical investigations furnished an opportunity to study the course of intra-annual and, partly, long-term dynamics of some of landscape indices. The studies revealed their slight spatial variability (within an elementary geosystem), which motivated the need for temporal observations with a large proportion of objectiveness. The dependence of the series of soil-geochemical indices on hydrothermal factors was emphasized. Time-coincident observations in doublet geosystems revealed substantial differences of soil-geochemical data obtained fro anthropogenically transformed geosystems, on parameters inherent in their natural counterparts [38].

Possibilities were demonstrated for using the synthesis of data from landscape-geochemical investigations in order to gain insight into the regularities of differentiation, integration and development of natural entities. Methods were developed for landscape-geochemical prediction and assessment of geosystems [39].

Station-based investigations of the behavior of substance in conditionally natural and anthropogenically modified geosystems that have been carried in different regions of Siberia, together with landscape-geochemical mapping and experimental research made it possible to create the technique for landscape-geochemical substantiation of regional systems of nature management. It includes such pivotal elements as diagnostics of the state of geosystems by analyzing long-term series of observations of migration of material, determining the stability of natural entities to technogenic impacts, and establishing allowable technogenic loads, predicting the behavior of material on the basis of temporal prolongation of experimental data, and spatial extrapolation of results obtained at a local level, with due regard for restrictive ecological factors.

Thus the emergence of the science of geosystems, the development of the principles of geosystem approach, and its use in tackling ecological-geographical problems, together with the landscape-geochemical approach, betoken a constant quest and the emergence of new ideas in geographical science. The system paradigm has given us the chance to revise the logical foundations of the science of landscape environment and draw a clear distinction between problems in physiography and sectoral geographical disciplines [3]. The science of geosystems has become the "core" of physiography that evolved into a modern science treating of natural systems and taking full advantage of the system approach, while sectoral sciences, primarily so-called "through" sciences: landscape geophysics, and landscape geochemistry, based on exploiting its theoretical base and its own quantitative methods, allows an understanding of the functioning of natural entities.

References

1. Sochava V.B. Definition of some notions and terms of physiography. Doklady In-ta geogr. Sibiri i Dalnego Vostoka, 1963, issue 3.
2. Sochava V.B. Geotopology as a branch of the science of geosystems. In: Topological Aspects of the Science of Geosystems. Novosibirsk: Nauka, 1974.
3. Sochava V.B. An Introduction to the Science of Geosystems. Novosibirsk: Nauka, 1978.
4. Sochava V.B. Structural-dynamic landscape science and geographical problems of the future. Doklady In-ta geogr. Sibiri i Dalnego Vostoka, 1967, issue 16.
5. Sochava V.B. Geography and Ecology. Leningrad, 1970.
6. Sochava V.B. Science of Geosystems. Novosibirsk: Nauka, 1975.
7. Sochava V.B. The system paradigm in geography. Izvestiya VGO, 1973, v.105, issue 5.
8. Sochava V.B. Geographie und Ökologie. Petermanns Geographische Mitteilungen, 1972, Heft 5.
9. Sochava V.B. Das Systemparadigma in der Geographie. Petermanns Geographische Mitteilungen, 1974, Heft 3.
10. Topology of Steppe Geosystems. Leningrad: Nauka, 1970.
11. Sochava V.B., Volkova V.G., Druzhinina N.P. et al. The method of comprehensive ordination in landscape science and biogeocenology. Doklady In-ta geogr. Sibiri i Dalnego Vostoka, 1967, issue 14.
12. The Study of Steppe Geosystems Over Time. Novosibirsk: Nauka, 1976.
13. Natural Regimes of the Steppes of the Minusinskaya Depression (in the Example of the Koibalskaya Steppe). Novosibirsk: Nauka, 1976.
14. Geosystems of the Piedmonts of the Western Sayan. Novosibirsk: Nauka, 1979.
15. The Southern Taiga of the Angara Region. Structure and Natural Regimes of Southern-Taiga Landscape. Leningrad: Nauka, 1969.
16. Natural Regimes and Topogeosystems of the Angara Region's Taiga. Novosibirsk: Nauka, 1975.
17. The Southern Taiga of the Irtysh Region (Experience of a Station-Based Investigation of the Southern-Taiga Topogeosystems). Novosibirsk: Nauka, 1975.
18. The Structure and Functioning of the Southern-Taiga Geosystems in the Irtysh Region. Novosibirsk: Nauka, 1982.
19. Nature of the Taiga Irtysh Region. Novosibirsk: Nauka, 1987.
20. Natural Regimes of the Middle Taiga of West Siberia. Novosibirsk: Nauka, 1977.
21. Mikheev V.S. The Upper-Chara Depression: Experience of a Topological Study of Landscape. Novosibirsk: Nauka, 1974.
22. Krauklis A.A. Problems in Experimental Landscape Science. Novosibirsk: Nauka, 1979.
23. Ryumin V.V. Dynamics and Evolution of South-Siberian geosystems. Novosibirsk: Nauka, 1988.
24. Sochava V.B. Theoretical background of mapping of the habitat. Doklady In-ta geogr. Sibiri i Dalnego Vostoka, 1973, issue 40.
25. Landscapes of the South of East Siberia, Map. M 1 : 1 500 000 / V.S. Mikheev and V.A. Ryashin. Ed. by V.B. Sochava. Moscow: GUGK, 1977.
26. Sochava V.B. Geographical forecasting as the most important direction of modern geography. Doklady In-ta geogr. Sibiri i Dalnego Vostoka, 1974, issue 43.
27. Isachenko A.G. Optimization of the Natural Environment. Moscow: Mysl, 1980.
28. Mikheev V.S. Landscape geographic support for Siberia`s complex problems. – Novosibirsk: Nauka, 1987.
29. Topological Aspects of the Science of Geosystems. Novosibirsk: Nauka, 1974.
30. Polynov B.B. Selected Works. Moscow: Izd-vo AN SSSR, 1956.
31. Glazovskaya M.A. The Geochemical Foundations of the Typology and Techniques for Investigating natural Landscapes. Moscow: Izd-vo MGU, 1964.
32. Perelman A.I. and Kasimov N.S. Landscape Geochemistry. Moscow: Astreya-2000, 1999.

33. Snytko V.A. Geochemical Investigations of Metabolism in geosystems. Novosibirsk: Nauka, 1978.
34. Snytko V.A. Substance Dynamics in geosystems. Geo Journal, 1983, No. 2.
35. Snytko V.A., Semenov Yu.M. and Suvorov E.G. Station-based landscape ecological research in Siberia. IALE Bulletin, 1995, v.13, No. 1.
36. Nechayeva E.G. Landscape-Geochemical Analysis of Geosystem Dynamics. Irkutsk, 1985.
37. Volkova V.G. and Davydova N.D. Technogenesis and Transformation of Landscapes. Novosibirsk: Nauka, 1987.
38. Snytko V.A., Semenov Yu.M. and Martynov A.V. Landscape-geochemical Analysis of KATEK Geosystems. Novosibirsk: Nauka, 1987.
39. Semenov Yu.M. Landscape-Geochemical Synthesis and Organization of Geosystems. Novosibirsk: Nauka, 1991.

CHAPTER

6

Landscape-geochemical Monitoring in the Baikal Region

E.G. NECHAEVA

The environment is regarded as a holistic and dynamic system (geosystem). Processes of interaction between the main landscape components (vegetation, soils, and water) govern accumulative and migration characteristics of geosystems. This mechanism ensures the structural tolerance of a landscape and carries some information on the trends of geosystems in the course of natural or anthropogenic evolution.

A program of landscape-geochemical monitoring is developed on the principles of functional organization of geosystems. A monitoring database includes contents of chemical elements in the environmental components and calculated relative characteristics of the self-purification ability of a geosystem. A landscape-geochemical map of a representative fragment of a given territory is laid out with regard to spatial-temporal dynamics of substances and results of analysis of environmental characteristics and anthropogenic loads. Differentiation of a territory by the degree of chemical pollution danger serves as an integral characteristic of the situation. In this paper, experience of drawing landscape-geochemical map for the mountain-taiga landscape of the southern part of the Baikal region is presented. Specific features of the monitoring program for plain regions of the humid zone are pointed out. In general, this study contributes to a global program of preservation of the biosphere.

Key words: water migration, self-purification ability, geosystem, mapping.

The system approach to studying the natural environment is typical of modern science. Within this approach, the environment may be treated as a holistic and dynamic system (geosystem) [1]. The study of the material structure of geosystems or landscapes has become a crucial line of geography. Acceptance of the dynamics as an integral form of landscape functioning makes it important to consider migration and accumulative processes that ensure the balance of substances and, hence, the structural tolerance of the entire natural complex.

The dynamical equilibrium state of geosystems is controlled by their invariant properties, i.e. by those properties which remain virtually constant under external impacts. The functional approach to studying the dynamical equilibrium state serves as a methodological basis for environmental monitoring. The informative characteristics of inter-component relations in the cyclic substance exchange deserve special attention in the development of a program of landscape-geochemical monitoring.

Mechanisms of interaction between plants and soils involve active participation of water that imparts sustainability to a geosystem dynamics and, at the same time, predetermines tendencies of its irreversible changes over the course of natural or anthropogenic evolution. The concepts of sustainable dynamics and directed development of geosystems provide the basis for the assessment of geosystem tolerance to different impacts and for the choice of proper ways to control the development of adverse phenomena.

Characteristics of the balance of substances in a given geosystem, its ecological quality, parameters of self-purification ability and, hence, the potential geosystem sustainability, and also indicators of their transformation are governed by the spatial differentiation. That is why, at the stage of landscape-geochemical survey, the main purpose of investigations implies an adequate choice of one or several regions representative in terms of landscape dynamics for subsequent monitoring. In essence, it is the first principle of the monitoring program.

Ecological landscape-geochemical zoning of large regions with regard to their natural diversity, the specific character specifics of industrial use, and distribution of pollution sources dictates the character of geographical work. The end target is to create territorial schemes of rational nature management, develop some models of rehabilitation, conservation, and optimization of landscapes realizing the partnership strategy between humans and nature [1]. This strategy implies public outreach concerning the properties of the biogeochemical environment, and creation and support of its sustainability favorable for humans and other habitants of an ecosystem.

To accomplish the tasks of landscape-geochemical monitoring of the state of undisturbed and disturbed environments requires a know ledse of the mechanisms of self-regulated processes of production, accumulation, transformation, and migration of substances. The character of these processes is determined by bioclimatic and geological-geomorphological factors. Their participation in the formation of the landscape-geochemical situation on a territory shows up in directions of geochemical fluxes of substances, their leaching from soils in eluvial positions, and, on the other hand, accumulation of oxidized compounds in contrastive reduction-oxidation conditions of the soil medium, etc. The reaction of soil medium (pH) is of particular importance. Its spatial heterogeneity is represented by the following gradation of acid-alkaline conditions denoted by the symbols of hydrogen and calcium: [H], [H, H-Ca], [H-Ca], [H-Ca, Ca], [Ca, H-Ca], [Ca]. Each class is characterized by certain values of pH from less than 4 to more than 8. Acid-alkaline parameters of the soil cover dictate the specificity of geochemical barriers; studying them is one of the major tasks of the landscape-geochemical analysis of a territory [2].

The second principle of the program is the multiple mapping of ecological conditions of a landscape medium and its differentiation by the degree of chemical pollution hazard. This stage of investigation is illustrated by a large-scale map of the southern part of the Baikal region (Figure and Table). It is the southern part that was chosen for mapping, because it calls the most for environmental control. In the analysis of the landscape geochemical environment, particular emphasis was placed on the conditions of substance migration (SM). Characteristics of relief and altitude of a territory were taken as criteria of SM intensity. Slight SM is inherent in foothill plains, 400–700 m; moderate – in low mountains, high plateaux, and foothills, 600–900 m; moderately high – hill sides, 600–900; high – medium mountains, high plateau, and hill sides, 800–1200; very high – steep hill sides, 800–1200 m; intensive – medium and high mountains, 1000–1700 m; very intensive – high mountains, 1600–2200 m a.s.l.

The territorial aspect of the program of landscape-geochemical monitoring makes it similar to the concept of landscape monitoring [3]. In the course of mapping of material dynamics in study regions, analytical data of diagnostics of the natural geochemical background as well as of the specificity of industrial waste, and of the direction and intensity of technogenic fluxes of matter should be collected.

Landscapes of moderately cold environment with limitedly developed vegetation and substance migration (SM): (1) moderate, (2) moderately high, (3) high, (4) very high; with optimally developed vegetation and SM: (5) moderately high, (6) very high. Landscapes of cold and very cold environment with limitedly developed vegetation and SM: (7) very high, (8) intensive, (9) very intensive; (10) with reduced vegetation and intensive SM; (11) anthropogenically disturbed landscapes with slight SM.

Geochemical classes of landscapes: (12) acid [H], (13) oxide-iron and acid [O-Fe, H], (14) the same with participation of transitional to calcium class [O-Fe, H, H-Ca], (15) acid and transitional to calcium [H, H-Ca], (16) transitional to calcium [H-Ca], (17) cryogenic organic-lithogenic, acid. (I–XXX) landscape units numbers.

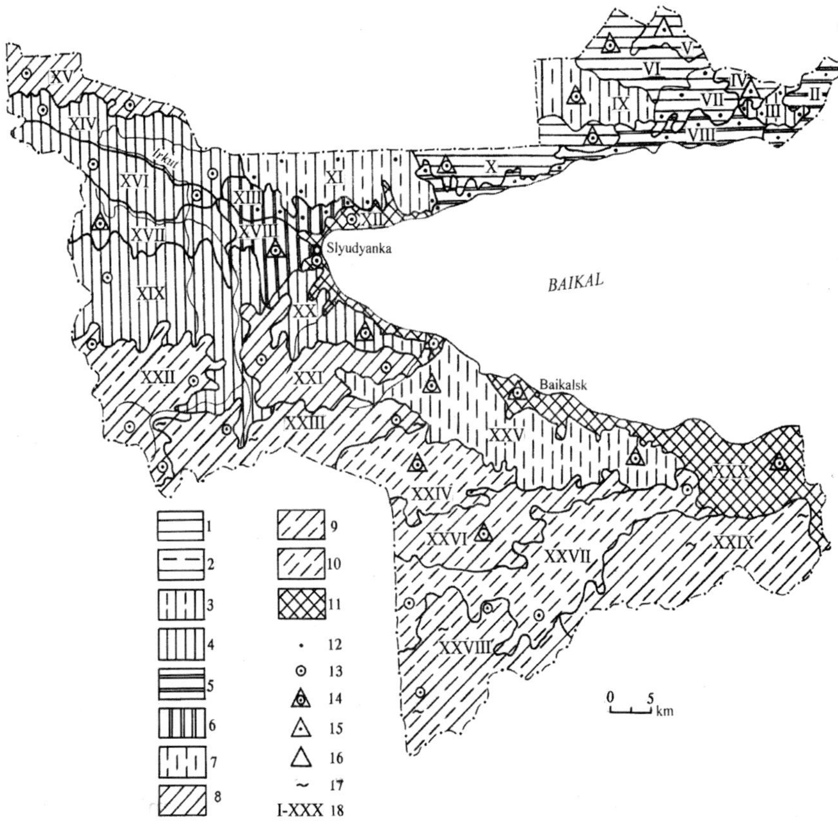

Figure: Differentiation of the ecological landscape-geochemical situation in the south of the Baikal region (territory of the Slyudyansky rayon, Irkutskaya oblast).

Points for regime observation of characteristics of the main landscape-forming processes are chosen on a cartographic basis. Taking into account the great variability of constituents of the medium, primary emphasis is placed on those parameters which reflect the essence of migration processes. In landscapes of the humid zone, these are the parameters of the carbon–calcium complex (the bicarbonate–calcium composition of natural water in the forest zone). In the arid zone, the composition of soluble salts is a more important indicator. Swamping processes are indicative of the dynamics of lowland-taiga landscapes, whereas water erosion is more significant for mountain-taiga landscapes.

Thus, monitoring programs, although based on common universal approaches, must be geographically adapted according to the substantial structure of landscapes and possible consequences of anthropogenic loads.

Table

Characterization of landscape medium on the south of the Baikal region

Altitude, m	Bioclimatic characteristics of landscapes*	Soils of landscape units (from I to XXX)	Anthropogenic loads and danger of technogenic chemical pollution
400–700	Moderate-cold (800–1200°)**, high- and moderate-productive (4000–8000 kg/ha)	Taiga podburic soils (PBt) in combination with illuvial-humus podzolic (Pih), soddy slight-podzolic (SPs), and valley swamped; in places saturated soddy brown (SBs) soils /XII, XXX/	Very high
600–1200	The same	Soddy podzolic in combination with PBt, soddy lithogenic (Slt), in places permafrost-taiga podzolized (PTpd) /III, VII, XI/ with participation of saturated soddy-taiga /VI, IX, X/, soddy-podzolic detrital-carbonate and soddy-carbonate podzolized (Scpd) /I, V/	High
800–1200	The same	SPs and Slt /II, VIII, XIII/, PBt in combination with Pih, PTpd /XIV, XVI, XIX/, and SBs /XVII, XVIII, XX/, in places Scpd /IV/	Moderate
1000–1700	Cold (600–800°), moderate- and low-productive (2000–6000 kg/ha)	PBt, Pih, Slt /XV, XXI, XXVII/ in combination with SBs and acid soddy-burozemic (Sba) /XXIV, XXV, XXVI/	Slight
1600–2200	Very cold (< 600°), extremely low-productive (<2000 kg/ha)	Mountain-meadow, mountain-tundra, and primitive mountain in combination with tundra podburic /XXIII, XXVIII, XXIX/, PBt and Sba in combination with Pih and Slt /XXII/	Very slight

* From Correlation Ecologic-Phytocenosis Map [7].
** Total average daily air temperature above 10 ° C.

The biogeochemical situation dictated by the contents of elements in a habitat is assessed by the criteria of landscape-dynamical condition of the region and by the substantive composition of landscape components. By means of special formulas, this situation is shown on a map displaying spatial differentiation of environmental conditions and factors. The numerator in every such formula contains elements, which are in excess for animals and humans within a given territory, and the denominator indicates deficient elements. The contents of formulas may be observed actually or calculated as a forecast from values of acid-alkaline and oxidation-reduction parameters, as well as from the specificity of technogenic chemical loads.

A monitoring data base serves not only for determination of the contents of individual chemical elements in a given habitat but also for calculation of relative characteristicsof the self-purification ability of the geo system via water migration and biogenic processes [4]. Integral functional characteristics of the intensity of transformation of organic compounds, biological cycling, water migration of particular elements, and other indices show the degree of tolerance and trends of possible changes in geosystems. The abovementioned and other characteristics of spatial-temporal structural-functional self-organization of the landscape sphere are useful for determination of permissible loads on a given geosystem, development of up-to-date resource-saving and nature-protecting technologies, and for other geographical application purposes.

A comprehensive geochemical analysis of the landscape sphere makes it possible to judge about the particular problems facing humans in a given environment. For example, in the course of geological exploration in the vicinity of the Kovykta gas-condensate field (GCF) on the Lena-Angara plateau, great disturbances of the hydrothermal regime of soils were found; besides, water-protective functions of local geosystems were also disturbed. Tree felling around drilling sites on hill sides resulted in the development of water-erosion processes and transformation of soil flow into surface runoff. Surface water is subject to pollution by readily soluble salts used in drilling operations. Also, salt pollution is caused by emergency ejections of head water enriched in calcium chlorides and a number of chemical elements unnatural for taiga landscapes. Upon drilling, calcareous rock is extracted to the surface and becomes the source of alkalization of surface water around drilling sites. A disturbance of the naturally established acid-alkaline equilibrium leads to changes in conditions of functioning of the accumulative-migration system in taiga landscape. The response of the medium shifts toward the alkaline part of the pH range, which results in decreasing mobility of lead and its accumulation in the landscape. In hydromorphic conditions, cadmium, copper, zinc, and cobalt accumulate in addition to lead [5].

In a thick moss-litter layer typical of taiga geosystems, lead shows high ability for accumulation. The value of the coefficient of biological absorption (Cb) defined as the ratio of the element content in that layer to its content in the soil varies from 1 to 22 in different places of the GCF region averaging about 8. These values of Cb for lead are only slightly below those for calcium (from 2 to 27; 9 on average). The third largest Cb (from 1 to 20; 6 in average) belongs to manganese. We take the sum of Cb of these three most active elements of the biological cycling as a characteristic of its intensity, which varies from 5 to 80 (25 in average) in the considered region. The value of the overall Cb and the contribution of different elements to it can be used to assess the character of changes in a landscape.

Analysis of the situation in the vicinity of the Kovykta GCF implies that a monitoring program for a taiga geosystem must include regime observations of the characteristics of basic landscape-geochemical conditions and processes of substance migration. In particular, data on the three abovementioned active elements of biological cycling and ion composition of water media are very important, especially in areas polluted by aggressive salts.

Background river and soil water of the region under study is characterized by bicarbonate-calcium composition and mineralization 0,2–0,3 g/l. Rainfall –

diluted water solutions of settling tanks on "worked out" drilling sites have a sulfate-chloride-sodium composition and mineralization of up to 3 g/l. Mineralization of the liquid phase of an emergency ejection of deep brine solutions reaches 700 g/l, more than one half represented by chlorine. Such elements-pollutants should receive primary attention with maximal consideration for the particular areas of their distribution and migration pathways.

Thus, in every particular situation, it is necessary to arrange monitoring programs with due regard for the character of technogenic loads and possible development of ecologically dangerous phenomena. Their prediction, prevention, and control constitute the main purpose of monitoring.

The study of environmental quality and main factors (hydrothermic, oxidation-reduction, acid-alkaline, and biochemical) of metabolism of substances in the areas of local anthropogenic loads should be accompanied by observation of substance arrival at a landscape with flows of interregional transfer. These data may introduce significant readjustment to the background geochemical characteristics, which are to be corrected periodically. Such an observation and, in general, the calculation of substances introduced are performed on the basis of data on chemical composition of the snow cover. Along with soluble salts, solid constituents settling in a litter layer on the soil surface and in the upper organomineral horizon upon snow melting should be taken into consideration.

The solubility of suspended substances increases significantly in the zones of influence of large chemical enterprises, because their aerosol waste markedly acidifies snowmelt water [6]. Areas with background contents of elements (with regard to global pollution) and locally polluted regions are distinguished rather exactly by the chemical composition of the snow cover. By early spring, at the time of maximal accumulation of substances in the snow cover, this represents a perfect object for observation of annual dynamics of technogenic pollution.

An equally effective object for monitoring a territory is provided by moss-bog areas presenting pronounced sorption-biogeochemical barriers. The substantial composition of moss-peat formations should be interpreted with due regard for their natural ability to selectively accumulate in the organic-hydromorphic medium of bogs certain chemical elements and increase solubility and mobility of others, such as silicon, aluminum, lead, iron, and a number of heavy metals belonging in the iron group. In contrast, bio- and water-migration characteristics of alkaline-earth elements in peat bogs are much lower than in automorphic soils of taiga landscapes [4].

Manganese is recognized as a typical taiga element, which is easily involved into biological cycling and accumulates in forest litter. In the southern taiga of East Siberia, the content of manganese in litter ranges from 2 to 10 g/kg ash substances (4.3 g/kg in average). Swamping sharply reduces the participation of manganese in biological cycling, and its content in the ash of eutrophic-peat layer comprises from 0.3 to 1.8 g/kg (0.8 g/kg on average). The

Cb of manganese in forest ecosystems of the region is about 40, while in oligotrophic-bog ecosystems it is about 2.

A typomorphic element of forest-bog geosystems, iron, shows an opposite behavior. Its biogenic and water-migration activity is an order of magnitude lower than that of manganese. Besides, as distinct from manganese, Cb of iron does not decrease but increases (from 0.3 to 0.9) upon swamping of forests. Forest litter contains from 10 to 40 g/kg ash (20 g/kg on average) of iron objects and the peat layer – from 30 to 70 g/kg (40 g/kg on average). The ratio of iron to manganese in bog geosystems (30–80) is an order of magnitude higher than that in forest geosystems (4–6). Thus the iron-to-manganese ratio can be used as a quantitative indicator of dynamic stages of swamping of taiga geosystems.

It should be noted that a knowledge of natural properties of a substance and specifics of its metabolism in different landscape conditions is necessary for adequate assessment of analytical data obtained and for prediction of a further development of an ecological situation (and, hence, for elaborating management strategies). For example, when we found notable accumulation of lead in peat-moss mass and green-moss cover of taiga geosystems in rather "clean" regions of Siberia, we explained this phenomenon by natural accumulation of the elements by mosses and not by technogenic loads. The participation of technogenesis in the accumulation of elements is detected only when information on background levels of their contents in such natural components as the biota, soil, and water is available. These levels can be determined as most frequently found element concentrations in a region (that might differ from clarke values).

For meadow-floodplain herbage in the limits of the taiga zone of the vast Ob'-Irtysh valley, the following contents of elements in 1 kg of dry substance are taken as natural standards: silicon, 3–37 g, calcium 3–17 g, magnesium 1–5 g, iron 0.1–0.3 g, titanium and manganese 0.1–0.2 g, barium 40–90 mg, strontium 20–100 mg, copper 7–10 mg, zirconium 5–12 mg, nickel 3–7 mg, vanadium 2–4 mg, chrome 1–3 mg, lead 1–2 mg, and cobalt 0.3–0.7 mg. Higher concentrations indicate pollution of the environment. Records of the temporal dynamics of a substance and a subsequent statistical calculation suggest a conclusion on for a particular situation.

River valleys deserve special attention in the assessment of landscapes as accumulative-migration systems. River valleys, located in the lowest places, receive surface and soil flows with natural and technogenic substances. Also, a specific accumulative process – bioproduction and sedimentation – develops in valleys. River floodplains undertake the harm of technogenic substances and, at the same time, owing to the water migration activity, perform the self-purification functions of a landscape.

River valleys provide attractive locations for settling and characterize the biogeochemical medium of a territory. In valleys, water and biotical components of survival, being basic and most susceptible to various impacts, serve as objects of monitoring. During floods, the substances entering the river network with industrial and agricultural waste come in contact with vegetation of floodplain ecosystems, accumulate in suspensions, and precipitate onto the soil surface. In long-flooded valleys of the Irtysh and Ob′, recently precipitated alluvial sediments contain per kg of dry substance 300–400 g of silicon, 50–100 g of aluminum, 20–50 g of iron, 6–13 g of calcium, 3–11 g of magnesium, 3–7 g of titanium, 0.4–0.5 g of manganese, 0.3–0.8 g of barium, 0.1–0.4 g of strontium and zirconium, 0.1–0.2 g of chrome, 40–130 mg of vanadium, 20–60 mg of nickel, 10–20 mg of lead, and 5–20 mg of cobalt (27 samples were analyzed).

The elements introduced, including heavy and toxic metals, accumulate in alluvial soils, and actively developing herbage partially involves them into biological cycling. The active use of polluted herbage for hay involves the inclusion of excess quantities of some elements to the trophoic chain, which may result in regional endemic diseases. In turn, they may superimpose on existing biogeochemical problems related to a deficiency of some elements, calcium in taiga landscapes, in particular.

The properties of environment dictate the quality of regional agricultural products, which, together with the hydrochemical situation affect human health. The assessment of biogeochemical conditions of living is one of the targets of the monitoring program. An integral characterization of the tolerance of landscape material structure to external impacts is another important target. This characterization can be based on a scale of landscape-geochemical parameters. The lowest points on the scale give grounds to forbid the use of geosystems for farming purposes, while high points, conversely, permit their use under proper restrictions.

It is an urgent challenge to implement the basic principles of the program of landscape-geochemical monitoring in the ecological evaluation of different nature-use projects, environmental control activity aimed at restoration of regenerative resources, design of cultural natural-anthropogenic landscapes favorable for human life in industrial regions, solving medical-geographical, agrogeochemical, and other problems. In general, the whole package of such measures should ensure an appropriate role of Russia in the global program of preservation of the biosphere.

REFERENCES

1. Sochava V.B. 1978. Introduction to the Theory of Geosystems. Nauka. Novosibirsk.
2. Perelman A.I. 1989. Geochemistry. Vysshaya Shkola. Moscow.
3. Semenov Yu.M. and Suvorov Ye.G. 1994. Toward the development of principles of landscape monitoring. Geography and Natural Resources 4.
4. Nechaeva E.G. 1994. Theory and experience of landscape-geochemical monitoring of the environment in plain taiga regions. Problems of Regional Ecology 3.
5. Glazovskaya M.A. 1988. Geochemistry of Natural and Technogenic Landscapes of the USSR. Vysshaya Shkola. Moscow.
6. Nechaeva E.G. and Makarov S.A. 1996. Snow cover as an object of regional monitoring of the environment. Geography and Natural Resources 2.
7. Ecological-Phytocenosis Complexes of the Asian Part of Russia. 1977. Irkutsk.

CHAPTER

7

Cartographic Study of the Biota in Southern East Siberia

A.V. BELOV, V.F. LYAMKIN AND L.P. SOKOLOVA

Southern East Siberia, stretching from the Yenisei river eastwards to northeastern Siberia and the (Russian) Far East, constitutes a complex geographic space both in terms of its genesis and current landscape appearance. This region includes a large portion of the Middle-Siberian upland, southern-Siberian mountains, southern Central-Yakutia plain, the near-Lena plateau, mountain systems of the Baikal-Patoma highland, and the Transbaikalia. Overall, the region has many similarities, especially in terms of its landscape climatic patterns, as an intra-continental region of the North-Asian subcontinent.

Accordingly, the biota (flora and fauna) in southern East Siberia is fairly inhomogeneous and constitutes a complicated heterogenous and heterochronous entity. It possesses a specific spatial geographic structure whose formation and current appearance are closely related to and mutually determined by the features of landscape climatic situation, both within the historic and present contexts. It is where run the boundaries of five large natural and, accordingly, biogeographical, regions [1, 2].

Naturally, this is reflected in the regional characteristics of how the vegetation cover responds to various forms of anthropogenic impact, entailing different approaches to solving problems of protecting the vegetation and animal population and of rational utilization of biotic resources.

Serious anthropogenic changes in vegetation cover and fauna in southern East Siberia, the result of many years of extensive exploitation of the region's floral, faunal and land resources, have exacerbated rendered the biodiversity conservation issue, as well as in many other regions of the country [3, 4, 5, 6].

The study and conservation of biodiversity is a comprehensive problem in basic and applied research. The various aspects of biodiversity were addressed by many scientists, both abroad [7, 8] and in this country [9, 10, 11].

On the other hand, there remain a great many poorly understood general theoretical and methodological issues relating to the problem of identifying biodiversity and its conservation, especially in the context of analyzing the structural-dynamical aspects of biotic organization. Meanwhile fundamental problems of studying the biotic organization (vegetation, and animal population) and biodiversity conservation at all levels of its manifestation are closely interlinked.

In accordance with adopted international documents, the term "biodiversity of a territory" includes a total intraspecific diversity of plants and animals, diversity of their species, as well as the diversity of ecosystems and natural complexes (geosystems), with the biota is an integral part of them. This extends considerably the validity range of the biodiversity concept to the level of biogeocenotic cover of a territory in all its manifestations.

For that reason, currently addressing the problem of biodiversity conservation no longer reduces to adopting prohibitive nature conservancy measures aimed at conservation not only of rare and endangered species. The Rio de Janeiro UN Conference (1992) elaborated an essentially novel paradigm of nature management that implies securing a balance between the resolution of socio-economic problems of a region and conservation of its natural environment, and satisfying the key vital requirements of the present generation of its inhabitants, and future generations.

At the present time the issue of biodiversity conservation has been brought to a level of addressing tasks of managing a territory's sustainable socio-economic development. In this case the ecological-functional, resource, ecological-economic, social, ecological-indicative and other aspects of biodiversity are becoming of decisive importance for tackling a great many issues related to rational nature management as an element of the region's sustainable development.

This approach to nature management can only be implemented on the basis of an in-depth scientific knowledge of the principles of the spontaneous and anthropogenic organization and functioning of the biosphere (or of the landscape sphere) and its components (vegetation and animal population) at all levels of its dimensions and organization, especially on a topological and regional level. It is important to identify the key evolutionary-genetic and factorial-dynamical processes in the biota, and their mechanisms and directions that determine its current composition, structure and natural ecological potential.

A broad understanding of biodiversity directs us not only towards its biotic gene-pool aspect that is treated in the evolution and (or) gene-engineering perspective. It is necessary to define the structural-functional hierarchy of priorities in biodiversity itself, based on trophic links and other ecosystem interactions between vegetation and animal population. Of course, it is the vegetation that provides the key component in most ecosystems based on these trophic relations and links. Along with trophic and other biogeographicalal links and relations, it is responsible for the characteristics of biocenoses or biogeocoenoses and, hence, of geosystems.

Moreover, it is necessary to take into account the biota's other important environment-forming, environment-protecting and other ecological functions, such as water-protection and water-regulation, anti-erosion, permafrost-stabilizing and other functions. Social (recreational, medicinal, etc.) functions of plant communities have recently acquired an increasing higher value, given the specific composition and structure of many of them.

These varied ecological parameters of biodiversity assessment must be taken consideration when determining the "ecological value of significance" of particular biotic complexes. This provides a scientific framework for economic calculations in assessing biodiversity, and elaborating and adopting regional normative-legislative acts on the protection and efficient utilization of the existing biodiversity. Information about the ecological significance of biotic objects is also necessary for designing various economic projects. Such an approach is to a certain extent realized when developing the ideology of landscape planning of ecologically oriented land use [12].

All this provides evidence that when defining the priorities of biodiversity conservation issues to be addressed in any region, it is necessary to concentrate most attention on the preservation of the cenotic structure of vegetation as the condition of conservation of individual plant species, and also as a basis for the living environment of the animal population, its subunits (populations, zoocenoses), and separate representatives of the fauna.

In this context, biodiversity conservation is based on such issues as identification and protection of the territory's indigenous (climax) vegetation that has established in the course of historical development of the territory and matches most closely territory's landscape-climatic potential. On the one hand, the indigenous vegetation can be regarded as a basic reference level from which to trace all anthropogenic changes in plant communities. On the other hand, indigenous communities are indicative of the natural complexes' bioecological potential, and can therefore be employed in reconstructing the progressive or natural potential vegetation that is important in addressing many biodiversity issues [13]. In this context, the statement about the invariant of the biotic community, suggested by V.B. Sochava [14], is of great interest in tackling issues related to conservation of cenotic diversity of vegetation.

Such a treatment of the biodiversity concept makes it possible to examine from the systemic standpoint the biodiversity conservation ideology itself as well. It, i.e. conservation, does not merely imply a set of prohibitive nature conservancy measures in respect of some biological object. It is a consistent package of measures on the study and identification of qualitative and quantitative parameters related to basic units of biodiversity, determination of its spatial-geographical structure, assessment of the identified biodiversity in terms of its ecological significance and present status with the identification of spontaneous and anthropogenic dynamical trends in the biota, and, finally, on making (on this information basis) forecasts and recommendations for the utilization, regeneration and conservation of biodiversity for a territory.

Accordingly, this implies identifying three stages of a common information-research process related to biodiversity conservation: inventory, assessment, and forecasting-recommendation, which is combined by the common application focus on the preservation and maintenance of the identified biodiversity for a given territory.

In cenotic biodiversity studies, the greatest information content corresponds to different-scale geobotanical, zoogeographical and comprehensive (integral) biotic mapping that provides an accurate geographical model for cenotic, dynamical and ecotopological structure of the biota or its components: vegetation and animal population for the study area.

Analysis of accumulated international and national experience in cartographic studies of biodiversity [11] demonstrated that there is emerging a special problem-oriented area of thematic mapping, namely, mapping of biodiversity, in which an important role is attached to cenotic diversity maps for the biota, primarily geobotanical inventory maps.

Mapping as a methodological investigative approach is applicable to all stages of studying, assessing and forecasting biodiversity conservation. Mapping of the biota for purposes of biodiversity conservation can therefore be regarded as a unified array of consecutive actions – from the creation of inventory maps of biodiversity and cartographic assessment of its state and quality to the drawing up of proposals for its conservation according to its territorial individual.

Combining inventory, assessment and forecasting-recommendation maps of the biota into a single information-research cartographic system has a number of advantages. First of all, it implies a certain scientific and methodological standardization for all maps to be generated. The requirements of this standardization are defined by the general orientation of the entire system towards objectivizing the information available on maps, as it will eventually determine the constructibility and scientific reliability of forecasts and practical recommendations for conservation or improvement of the quality of the biota, and the natural environment as a whole.

This standardization also suggests mapping to comparable scales. It is advisable to maintain scale contingency of cartographic research for all map groups, from universal (inventory) to forecasting-recommendation ones. This measure provides for spatial and substantive concreteness and objectivity in all phases of cartographic research, especially – this is particularly true of the compilation of special-purpose assessment and forecasting-recommendation maps.

Within each stage of cartographic research of vegetation and animal population, it is advisable to do work at different scales – from a large scale to a reference-informational small scale. Such a different-scale character of cartographic research provides more detailed and more accurate information about the composition and structure of plant communities and zoocenoses, their linkages with ecotopes, and about the spontaneous and anthropogenic dynamics. Eventually all this improves objectivity of classification generalizations as reflections of multi-step and multifactor systems of organization of vegetation or (and) animal population and, as a consequence, supplies fuller information to a next stage of cartographic study of the biota.

The largely applied character of the entire cartographic research system focusing on optimization of environmental quality, in turn, determines the set of attributes and properties of vegetation and fauna which should be scrutinized when doing inventory work. More importantly, this would make researches focus on determination of their conceptual theoretical and methodological viewpoints in the study of the biota as the most important, integral dynamical component of the natural (geographical) environment [14].

This shows that all stages require an independent solution of fundamental scientific-theoretical and methodological problems, as well as applied issues related to mapping of vegetation and animal population. Scientifically, the researcher's main concern should be with the identification of the biotic and geographical framework and principles of organization of vegetation and animal population as integral elements of natural systems (geosystems), with due regard for their genesis and high degree of spontaneous and anthropogenic dynamicity.

System (geosystems) approach to studying the biotic components' organization in natural complexes whose main theoretical postulates were laid down by V.B. Sochava [14], ensures high saturation of maps with varied ecological information. They contain information about the ecologically ecologically significant geographic factors that are responsible for the structure and dynamics of vegetation and animal population at different levels of their organization, about the dynamical and ecological potential of territories occupied by particular biotic taxons, and much other information.

As regards practical implications, investigations should be focused on an improvement of objectivity, geographical accuracy and constructability of the cartographic conclusions and recommendations made. The degree of their objectivity will depend directly on our understanding of the biota, the character of its organization, as well as on the functional role in natural systems and availability of the necessary information in all interrelated phases of a common cartographic investigative process.

Mapping of vegetation and animal population has a number of characteristic features because of the specific character of the study objects themselves. Geobotanical and zoogeographical cartographic studies of biodiversity could be carried out concurrently, yet complex synthetic maps relying on basic component maps may well be generated.

Geobotanical inventory maps, built on the principles of a structural-dynamical classification of vegetation, provide a valuable information basis for a special-purpose assessment of plant communities. Such cartographic efforts on the study of the structural-dynamical and cenotic characteristics of vegetation were mounted in the Biogeography Laboratory of the V.B. Sochava Institute of Geography SB RAS for the entire region of southern East Siberia [15] as well as for its different regions: in the Prebaikalia, in the Baikal region, in the Transbaikalia, and others [16]. It was possible to generate universal geobotanical maps to a scale of 1:2,500,000, 1:1, 500, 000, 1:500, 000, and 1:25,000, providing a rather thorough portrayal of the spatial-geographical characteristics of cenotic and dynamical organization of vegetation in the study areas.

The geographical-genetic approach to analyzing the modern cenotic structure we have used in these studies, calls for an understanding and consideration of systemic organization of vegetation. Each plant community carries a progressive or regressive evolutionary-dynamical tendency that determines its status in the structure of cenotic diversity of current vegetation. Some of such communities are vegetation relicts from the distant past, and they have been and are preserved by the specific conditions of modern habitats: microclimate, special soil-hydrological conditions, and permafrost or quite the reverse, a thermal spring. The other part of rare phytocenoses, especially in contact areas of communities of different types of vegetation, such as forest or steppe vegetation, or their independent phytocenogenetic groups such as the Ural-Siberian, Bering and Angaride fratries, are produced through the inclusion of cenoelements of contacting communities over the course of dynamical tendencies associated with long-term (secular) cycles of climate change. These factors must also be taken into consideration when working out target-oriented measures related to conservation of a particular plant community.

On the other hand, the resolution of the vegetation biodiversity issue must be based on taking into account its high spontaneous and anthropogenic dynamicity. These manifestations of dynamicity of plant communities are different for a different type of vegetation, in the taiga-forest and steppe vegetation, for example. Of critical importance here are the edificatory properties prevailing in plant communities which provide a means to transform the influence on plant communities exerted by external environmental factors or by anthropogenic impacts.

Thus, within the context of vegetation biodiversity conservation, it is necessary to have a large amount of information about the various attributes and properties of plant communities themselves, about their faunistic composition, resilience of cenotic structure, their geographical and territorial definiteness, as well as about the varied functional role they play in the region's vegetation in general. Further, of important significance is, of course, an assessment of the degree of uniqueness of a particular plant community, and identification of the genesis (natural or anthropogenic) of this feature inherent in it.

Estimation criteria for the value of plant communities for the purposes of biodiversity conservation could be represented by such characteristics as the present uniqueness or rarity of communities, the genesis of this uniqueness or rarity, the degree of their preservation, sensitivity to impact factors, the place and role in the spatial combination of plant communities, its ecological role (function) in a natural complex (geosystems), etc.

Of special significance for biodiversity conservation are the applied maps that display the degree of such an assessment of vegetation and fauna in the particular regions. It is known that estimating mapping, including geobotanical mapping, has rather long existed as an area of thematic mapping, yet it was given impetus to its further development because of the enhanced role played by assessment of the natural potential for territories and environmental quality conservation. Assessment studies are based on investigating the links and interactions between natural complexes (geosystems) and their biota and economic systems of different kinds and their elements.

Overall, cartographic assessment of the biota implies the following key lines of inquiry:

- assessment of the degree of disturbance (preservation) of composition and structure of vegetation and animal population;
- assessment of spontaneous variability (dynamicity) of the biota;
- assessment of the resource and (or) ecological potential of vegetation and animal population; and
- assessment of resilience of the biota to anthropogenic factors.

In making assessments of the degree of disturbance (preservation) of vegetation, it is customary to take into consideration any manifestations of the various changes in floristic and cenotic composition, and in the structure of vegetation that evolve mainly under the influence of anthropogenic factors. The degree of anthropogenic disturbance of vegetation is determined by criteria of deviation in composition and structure of plant communities from their native state or from a selected state for a given time. An assessment can be made both for the entire vegetation and for its particular categories, including for species populations that are of the greatest significance in the economic practice or have another value, such as in terms of biodiversity conservation.

Assessment of spontaneous variability (dynamicity) of vegetation is essential to the understanding of the mechanism of natural resistance of plant communities to fluctuation changes in natural and climatic conditions, as well as to gain insight into floristic and cenotic biodiversity of a territory. This is especially true in regard to steppe and meadow vegetation that responds most sensitively to cyclic changes in growth conditions.

Assessment of the resource and ecological potential of vegetation is critical to defining the methods and scales of economic activities in a region. As far as assessment of the resource potential of vegetation is concerned, a wealth of scientific and empirical experience in forestry and agriculture has been accumulated to date. There are also preliminary results derived from assessing the resource potential of vegetation with medical and recreational implications.

Assessment of the ecological potential of vegetation is a relatively new avenue of inquiry, based on current views of the multifunctional environment-forming and environment-protective role of plant communities in natural complexes, into the composition of which they enter. Cenotic and ecotopological analyses are used to identify groups of plant communities according to categories of environment-forming (biostationary, water-regulation and other) and environment-protective (anti-erosion, permafrost-stabilizing, water-protective, avalanche- and mudflow-protective and other) properties. Where necessary, it is customary to select preferential functions of plant communities, sustaining which is the task of rational nature management.

Cartographic assessments of the ecological-functional role of the vegetation in the south of East Siberia were carried out by researchers of the Biogeography Laboratory of the V.B. Sochava Institute of Geography SB RAS [17-19]. It was shown that plant communities as a component of geosystems perform quite varied environment-forming and environment-protective functions that determine the ecological value of these communities as regards conservation of environmental quality and phytocenotic diversity.

They must form a basis for working out territorial principles of regulating the economic activities in order to provide for cenotic diversity conservation of vegetation. Where appropriate, it is also important to implement measures for natural or artificial regeneration of disturbed communities. Overall, a system of relations must be engineered, which would take into consideration the ecological and socio-economic interests of society in the streamline toward sustainable development of the region.

These assessment studies of vegetation are all interrelated and complement one another. Yet, their efficiency is enhanced when they are built on universal geobotanical maps which provide large amounts of he necessary information for assessment.

Cartographic study of the anthropogenic disturbance of vegetation implies assessing the present state of its spatial-geographical and cenotic structure as well as identification of the basic dynamical tendencies inherent in it, and factors that determine their further development. Such studies are subdivided into two areas:

One involves compiling universal geobotanical maps to different scales, by principles of a structural-dynamical classification of vegetation [14]. Such an approach provides a means to take into consideration the main regional-typological characteristics of vegetation as ell as the basic dynamical processes occurring in it and caused by anthropogenic or natural factors.

Compiling such a geobotanical map sets the stage for geobotanical assessment of disturbances of vegetation. Structural-dynamical analysis is used to obtain a retrospective assessment of the existing vegetation to identify native (initial) and temporal (derivative, or secondary) communities, followed by combining them into progressive sequences or by placing them into a category of stably-derivative communities. `Invariants of epistructures for plant communities are determined thus inferring the basis (or zero) assessment level used as a reference from which current spontaneous and anthropogenic changes in vegetation cover are assessed. In this case it is possible to establish the direction and character of changes in vegetation under the influence of future anthropogenic effects.

The other line of inquiry involves compiling various applied assessment maps: assessment of disturbance of vegetation cover, assessment of resilience of vegetation to particular anthropogenic factors influencing it, etc. This research area is closely linked to the former area, as it relies on universal botanical maps. In additional to universal maps, sue is made of a large number of other cartographic products and published data to obtain the various parameters of disturbance of plant communities.

Such efforts relating to cartographic assessment of vegetation can be exemplified by the map "Vegetation disturbance in the Prebaikalia" at a scale of 1:2,500,000 that was compiled in the Biogeography Labotatory of the V.B. Sochava Institute of Geography SB RAS (Fig. 1). This map was generated on the basis of the existing new universal geobotanical map with due regard for the increasing tendencies toward destructive anthropogenic alterations to vegetation cover over extensive territories, and for the decline in activity of regeneration processes in disturbed communities. On the universal geobotanical map the derivative communities were typisized not within progressive sequences that were characteristic for the Prebaikalia's vegetation in the not too remote past, but within sequences of anthropogenic transformation for which disturbance criteria for initial native plant communities are crucial.

Work on the map involved, in addition to analyzing vegetation disturbances, studying a large number of other cartographic documents to gain insight into the disturbances of plant communities, such as the boundaries of ploughed fields and agricultural lands, forests damaged by entomic pests, burned-over areas, timber felling areas, and industrial impacts. Besides, the mapping effort used information about fires, rare and endangered plant species and plant communities, archival agricultural and forest inventory data, and much other information.

A special analysis and assessment of the state of plant communities on the map revealed five categories of vegetation disturbance: arbitrarily undisturbed, weakly-, moderately- and strongly disturbed, and destroyed vegetation (see Fig. 1). Arbitrarily undisturbed vegetation is represented mainly by native dark-coniferous as well as mature pine and larch forests, sub-barrens thin forests, mountain pine thickets, and by mountain-tundra communities. Weakly disturbed vegetation includes mature and over-mature derivative forests, and mountain-tundra communities. Moderately disturbed vegetation encompasses medium-age and young light-coniferous forests as well as birch and aspen forests, and sub-barrens thin forests in mining places and steppes. Strongly disturbed vegetation is represented by cutting and burned-over areas, forests damaged by entomic pests as well as by meadow-steppe areas that are intensively utilized for pasture purposes. Vegetation (both removed and destroyed) is displayed in places of present-day ploughed territories.

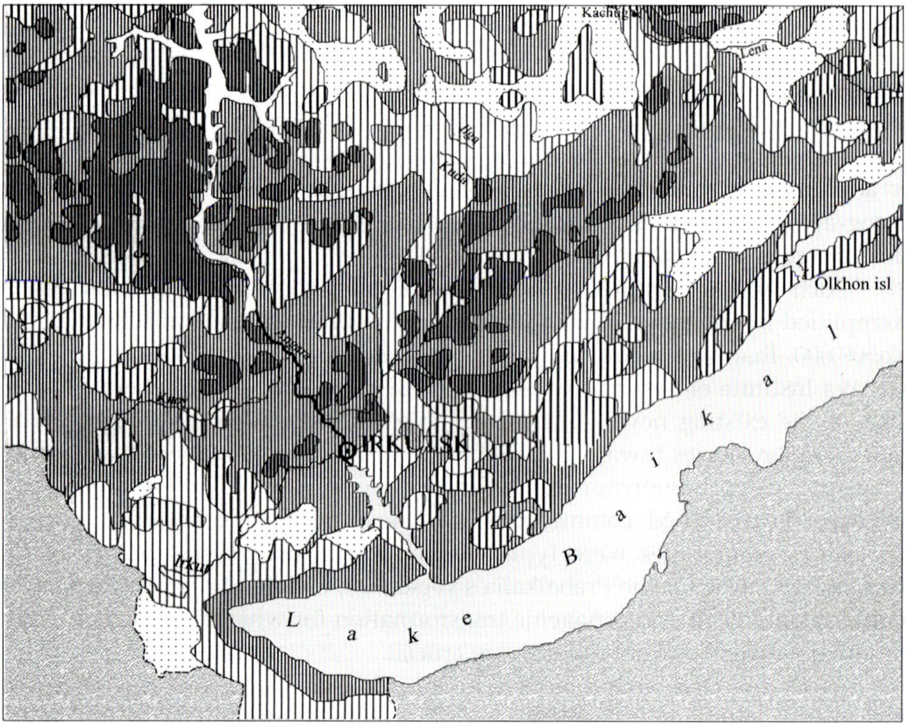

Degree of disturbance		Characteristic of vegetation
	Arbitrarily undisturbed	Native, mostly dark-coniferous, as well as mature pine and larch forests, sub-barrens thin forests, mountain pine thickets, and tundra and barrens communities. Disturbance factors: forest fires.
	Weakly disturbed	Mature and overmature pine and larch derivative forests and mountain-tundra communities. Disturbance factors: forest fires, and selective cutting.
	Moderately disturbed	Medium-age and young light-coniferous, as well as birch and aspen forests, sub-barrens thin forests and preserved steppes and meadows. Disturbance factors: forest fires, commercial cutting, and mining.
	Strongly disturbed	Cutting and burned-over areas, and also forests damaged by entomic pests, and meadow-steppe areas. Disturbance factors: intermittent forest fires, entomoparasites, and intense commercial cutting.
	Destroyed	Plowed lands. Disturbance factors: plowing of lands

Figure 1: Vegetation disturbance in the Prebaikalia (a portion of the 1:2 500 000 map).

Thus the maps thus compiled reflect two processes: anthropogenic destruction of vegetation under the influence of all factors that disturb vegetation structure and regeneration processes brought about by such anthropogenic impacts. Furthermore, there is a predominance of destructive processes accompanied by accumulation of attributes of a consistent decay of native vegetation over extensive territories. Naturally occurring restoration processes of vegetation no longer ensure resilience of native vegetation to anthropogenic impacts, increasingly extensive territories are becoming colonized by various derivative groups and, hence, the basic functions of vegetation are lost or attenuated, and the course of naturally occurring landscape process is disturbed.

According to the degree of disturbance of the vegetation cover, zones of territory development are clearly identified, where the economic activities should be correlated with the ecological potential of vegetation disturbance.

Thus, for the Prebaikalia we identified several such zones. The southwestern zone that encompasses highlands and medium-highlands of the Eastern Sayan mountains, is, for the time being, assigned to an arbitrarily undisturbed category. Yet, ecological hazards have recently presented themselves, as these highland territories are, to an increasingly greater extent, being developed by mining industries, with serious mechanical damage to the natural environment that fully destroys the biota in some places. Hence specially significant (water catchment, anti-avalanche, anti-erosion, permafrost-stabilizing) and other important functions of vegetation cover are disturbed.

The southern zone (heavily developed and strongly destroyed) lies along the Trans-Siberian Trunk Railroad. These areas are utilized for agricultural purposes (rangelands, ploughed fields, hayfields, gardening, and horticulture). Up to 80% of agricultural lands in the region lie along railroad lines, within a band to a distance of up to 50 km from them. Final timber felling areas are also concentrated here, and only mature tree stands are utilized for such purposes, which meet best in their quality the consumer requirements of the economy, in regard to 100-140-year-old pine, for example [23].

The middle and northern zones are moderately disturbed; they lie in areas of the Middle-Angara, Upper-Lena and Lower-Tunguska regions. In the short run these zones are likely to become strongly disturbed, as they include oil and gas bearing areas with the Kovykta, China, Tolokan and other gas and condensate fields, both discovered and prepared for development.

The northeastern zone that partially occupies the areas along the Lower Tunguska and the Patoma upland, is weakly disturbed at present. In the future this zone could undergo serious disturbances of vegetation cover and animal habitats, as it is of considerable interest for future economic development.

An independent direction of biota mapping in the interests of biodiversity conservation includes efforts related to compilation of inventory and assessment zoogeographical maps. Zoogeographical mapping, largely as an independent line of inquiry into animal population biodiversity for a territory, provides interesting results from different levels of study: population, and structural-cyclic. Such research efforts were made by the Biogeography Laboratory of the V.B. Sochava Institute of Geography SB RAS for various areas in southern East Siberia, focusing both on mammals [16, 20] and on a poorly understood group of invertebrates, the soil fauna [21]. All that research was done on a landscape-dynamical basis using information about structural-dynamical diversity of vegetation as a component of geosystems.

In the study of the animal population, much attention was also paid to the evolutionary-genetic background for the shaping of faunistic complexes in the study areas. Yet, because of the high spontaneous and anthropogenic dynamicity of the animal population: temporal (seasonal, over different years), territorial (stationary, migration), etc., of greater importance in the study of biodiversity was the inclusion of just these aspects of dynamics, the anthropogenic dynamics in particular. Hence it was possible to make assessments of the overall disturbance of ecological-faunistic complexes to identify different degrees of disturbance (weak, moderate, strong, and others) as well as to establish the categories of rarity or uniqueness for the particular animal species and justify the need for their protection on a federal or regional level. In this case the animal species thus identified were assessed in accordance with the relevant federal and regional Red Data Books of rare and endangered animals.

Different-scale (1:2, 500,00, 1:200,000, and 1:25,000) cartographic products were generated, based on studying biodiversity of animal population in various areas of southern East Siberia: the Prebaikalia, the intermontane depressions of the Northern Transbaikalia, the Lake Baikal watershed basin, and others [16, 20].

At the present time, in circumstances where the natural environment is undergoing dramatic restructuring, changes in vegetation cover give rise to an abrupt adverse change in habitats of many animal species. In this context, nature conservancy measures must be implemented, with special emphasis on the zoota of biocenoses, using geobotanical maps, and zoogeographical maps compiled on their basis.

Critical consequences of agricultural development of southern East Siberia for the fauna of land invertebrates include, in the first place, almost total destruction of steppe and meadow landscapes, the habitat of the steppe faunistic complex (steppes have persisted in the Olkhon area only). Secondly, the emergence (in the south of the region, and along the Trans-Siberian Railroad) of a large tract of open spaces, a serious ecological barrier in the path of exchange of genetic material between taiga animal populations in the near-Sayan and near-Angara areas, as a result of which they were dissected into two parts. This refers to hoofed animals: roe deer, Manchurian deer, wild boar, elk, musk deer, and reindeer.

In forests in the south of East Siberia, anthropogenic impacts and naturally occurring processes brought about dramatic changes that affected the structure of age and species composition, as well as the completeness and quality of forests. Currently virtually all landscapes in the surroundings of large human settlements at a radial distance of 50-60 km have become a recreational area, with the resulting abrupt decrease of the numbers of wild animals, primarily such large mammals as roe deer, fox, badger, and others, as well as land-nesting bird species (bunting, Tetraonidae, willow-warbler, and nightingale), and bird species nesting on bushes and low trees (thrush, and white-throat).

One further important factor influencing the biota is pasturing – its effects are both immediate (disturbances in nesting conditions, destruction of nests of land-nesting species) and indirect (modifications of the habitat cause meadow, steppe and, partially, forest habitats of animals to turn into overgrazing of a different degree of degradation. At the period of developing virgin and old idle lands the grazable lands decreased in their area, with the result of an abrupt increase in the intensity of load on them thus still further worsening the living conditions for steppe species of animals. In this context, grazable lands almost totally lack populations of land vertebrates.

Generation of assessment maps for the Prebaikalia's territory gives an insight into the role played by the zoota in the shaping of the ecological situation in the region as well as shows the characteristics of biological diversity and its present status. By disturbance or degradation of the fauna is meant here any changes in contemporaneous populations and communities which imply a population decline, a reduction and fragmentation of the areas, a change in species structure of communities, including the introduction of new species of wild animals, as well as a change in the habitat, as a result of which the existence of aboriginal species or their communities becomes impossible.

Such studies on disturbances of the fauna may be exemplified by the map "Fauna disturbance in the Prebaikalia" (1:2,500,000) (Fig. 2) which gives an idea of the present status of animal communities in this region. Because the chief cause for species' extinction is degradation of natural environment, much attention in devising the legend was paid precisely to the changes in animal habitats and, primarily, in vegetation cover, the environment-forming significance of which is universally known.

Plant communities were treated as the key component of biocomplexes that characterizes a particular complex of animals living in it. Where the vegetation has the greatest influence on the distribution of animals, it is, nevertheless, a good indicator of the entire biocomplex and clearly reflects its intrinsic natural characteristics and secondary changes. In this connection, the maps "Vegetation cover" and "Vegetation disturbance" [16] for the Prebaikalia, and other cartographic and published data provided the basis for the compilation of the map "Fauna disturbance in the Prebaikalia".

The map legend reflects three degrees of disturbance of aboriginal ecological-faunistic complexes and the ichthyofauna: weak – up to 30% of the area occupied, moderate – from 30 to 50%, and strong – 50-80%, as well as three degrees of disturbance (weak, moderate, and strong) of the ichthyofauna. Each number of the legend provides information about the exinct or critically threathened species. In addition, the causes for disturbances are indicated. The map displays zapovedniks, zakazniks and the national park, having a special significance for conservation and restoration of natural complexes, and for maintenance of an ecological equilibrium.

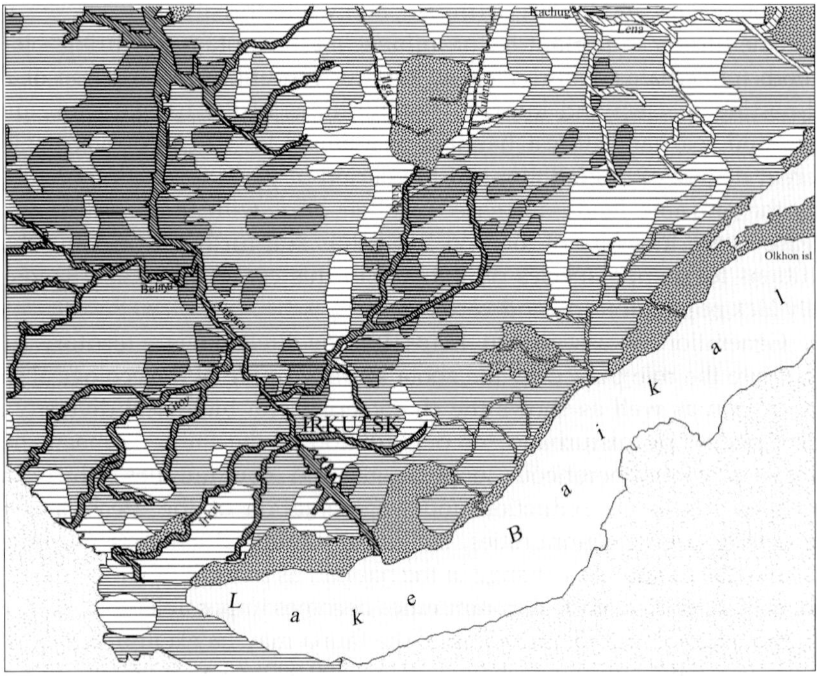

DISTURBANCE OF ABORIGINAL ECOLOGO-FAUNISTIC COMPLEXES

Zapovedniks, zakazniks, and national parks - very weak, less 30%.

Weak, up to 30% of the area occupied. Species in a critical state: Siberian goat, snow leopard, red wolf, black-capped marmot, gyrfalcon, peregrine, and saker falcon. Causes: forest fires, and technogenic factors.

Moderate, 30-50% of the area occupied. Species in a critical state: golden eagle, white-tailed and long-tailed bald eagles, swans, storks, cranes, Asian dowitcher, and osprey. Causes: overhunting, tree felling, forest fires, technogenic pollution, and intensive recreation.

Strong, 50-80% of the area occupied. Extinct species: bustard, and osprey. Species in a critical state: steppe eagle, imperial eagle, Daurian partridge, and white-tailed eagle. Causes: plowing of lands, tree felling, overgrazing, technogenic pollution, overhunting, and recreation.

DISTURBANCE OF THE ICHTHYOFAUNA

Weak: the regime is similar to a natural one, a reduction in population of separate fish species as a result of unorganized fishery. Local fishing.

Moderate: disturbed existuence regime. Species in a critical state: nelma, Lena sturgeon, taimen, lenok, round whitefish, common whitefish, and silver trout. Causes: overfishing, pollution from industrial and domestic effluents, and water transport.

Strong: fully disturbed existence regime. Extinct native forms of sturgeon and starlet. Species in a critical state: tugun, common whitefish, taimen, lenok, and grayling. Causes: disturbance of the hydrological regime due to the construction of hydroelectric stations, timber rafting, technogenic pollution of runoff, and commercial and local fishing.

In connection with acclimatization of sturgeon, ciscos, peled, omul, and bream, genetic contamination of aboriginal species is possible.

Figure 2: Fauna disturbance in the Prebaikalia (a portion of the 1:2 500 000 map).

The color pattern of the map purports to express different meanings. Zapovedniks, zakazniks, and the national park are displayed dark-green in color; bright-green corresponds to contours with a weak degree of disturbance. Orange color for moderately disturbed complexes implies a warning that the faunistic complexes on these territories require care and caution, otherwise disturbances are able to lead to a loss of habitats for various animal species or of spawning places for many fish species, as indicated by red color.

In addition to separate cartographic studies on flora and fauna diversity, the Biogeography Laboratory of the V.B.Sochava Institute of Geography SB RAS carried out comprehensive assessments of biodiversity for the Baikal natural territory. These research efforts encompassed the entire process of cartographic examination of the biota as an integrated program of its structural-dynamical and landscape-ecological analysis to define independent phases with a common goal of biodiversity conservation. The groundwork for this system was provided by universal (inventory) geobotanical and zoogeographical maps [15, 16].

As part of their special-purpose analysis, with due regard for the disturbance of the biota and its resilience to anthropogenic impacts, it was possible to identify biotic complexes of a different degree of ecological significance (value) – from high to low. This information serves as a basis for the cartographic expression of ecological zoning of the biota for the Baikal natural territory to define the principal goals of the utilization of the region's biodiversity which takes proper account of the need for its conservation and restoration [16].

Given the present conditions of the market economy, a successful resolution of biodiversity conservation issues must, obviously, be accompanied by integrated ecological-economic assessments of biodiversity itself as a natural resource, and results derived from such assessments would be reflected on specialized maps of land cadastre [22].

In closing, the research results reported here have demonstrated that in order to make correct scientifically-justified management decisions on various ecological-geographic problems arising in the economic development of southern East Siberia, including those related to the preservation of the region's vegetation biodiversity, requires doing comprehensive cartographic research into the biota (vegetation and animal population): a complete inventory of its composition, identification of the geographical and cenotic framework of its organization, assessment of its status and ecological-functional role, making predictions of its further development as well as recommendations for its conservation.

This research work was done with financial support from the Russian Foundation for Basic Research, project No. 04-05-64078.

References

1. Sochava V.B., Timofeev D.A. Physiographic areas in Asia. Transactions of the Institute of Geography of Siberia and the Far East. Irkutsk, 1968, issue 19, pp. 3-19.
2. Physiographic Regionalization of the USSR. Ed. by N.A. Gvozdetsky. Moscow: Izd-vo MGU, 1968, 574 p.
3. Malyshev L.I., Peshkova G.A. They are in Need of Protection: Rare and Endangered Plants of Central Siberia. Novosibirsk: Nauka, Sib. Division, 1979, 172 p.
4. Molozhnikov V.N. Plant Communities in the Baikal Region. Novosibirsk: Nauka. Sib. Division, 1986, 272 p..
5. Boykov T.G. Rare Plants and Phytocoenoses in the Transbaikalia. Novosibirsk: Nauka. Sib. Enterprise RAS, 1999, 264 p.
6. Belov A.V., Sokolova L.P. Mapping the vegetation in southern East Siberia within the system of the region's biodiversity conservation. Cartographic and Geoinformation Support of Regional Development. Management. Irkutsk: Institute of Geography SB RAS Publisher, 2002, pp. 27-31.
7. Woodwell G. Problems and Prospects for Biodiversity Conservation. Science, 1991.
8. Wittaker R. J, Willis K.J., Field R. Scale and Species Richness: towards a general, hierarchical theory of species diversity . J. of Biogeogr., 2001, v. 28, pp. 453-470.
9. Yurtsev B.A. Ecological-geographic structure of biological diversity, and the strategy for its inventory and protection. Biological Diversity: Approaches to the Study and Conservation, SPb, 1992, pp. 7-21.
10. Malyshev L.I. Biological diversity in the spatial perspective./ Biological Diversity: Approaches to the Study and Conservation, SPb, 1992, pp. 41-52.
11. Ogureeva G.N., Kotova T.V. Mapping biodiversity. Geography and Biodiversity Monitoring. Moscow: NYMTs, 2002.
12. Landscape Planning: Principles, Methods, European and Russian Experience. Bonn-Moscow-Irkutsk: Institute of Geography SB RAS Publisher, 2002, 141 p.
13. Ricotta C., Karranza M.L. Maps for potential natural vegetation as an alternative to computer generated neutral landscape models? In: Geobotanical Mapping: 2001- 2002. SPb, 2002, pp. 16-23.
14. Sochava V.B. Vegetation Cover on Thematic Maps. Novosibirsk: Nauka, 1979, 190 p.
15. Belov A.V. The map of vegetation in southern East Siberia: principles and compilation methods. In: Geobotanical Mapping 1973, Leningrad: Nauka,1973, pp. 16-30.
16. Belov A.V., Lyamkin V.F., Sokolova L.P. Cartographic Studies of the Biota. Irkutsk: Oblmashinform, 2002, 202 p.
17. Belov A.V. Plant protection and long-term prediction issues in tapping the vegetative resources in the middle and southern taiga in Siberia Transactions of the Institute of Geography of Siberia and the Far East. Irkutsk, 1975, issue 48, pp. 56-63.
18. Belov A.V., Lavrenko N.N. The map of vegetation and its ecological potential for the BAM area. Basic Sciences for the National Economy. Moscow: Nauka, 1990, pp. 512-513.
19. Belov A.V., Volkova V.G., Davydova N.G. Environment-protective properties of vegetation. Regional Ecological Atlas. Novosibirsk: Publishing House of SB RAS, 1998, pp. 98-104.
20. Lyamkin V.F. Zoogeography and ecology of mammals in intermontane depressions of the the Baikal rift zone. Irkutsk: Institute of Geography SB RAS Publisher, 2002, 133 p.
21. Bessolitsyna E.P. Landscape-Ecological Analysis of the Structure of Soil Zoocoenoses in southern Siberia. Irkutsk, 2001, 166 p.
22. Principles and methods of economic assessment of lands and wildlife: Analytical Handbook. Moscow: The GEF "Biodiversity Conservation" Project, Institute of Economics of Nature Management, 2000, 101 p.
23. Forests and Forestry of the Irkutsk Region. Vashchuk L.N., Popov L.V., Krasny N.M. et al. Irkutsk, 1997, 288 p.

PART III

SOCIO-ECONOMIC GEOGRAPHY

CHAPTER

8

Siberian Macro-Region in the Economy of Russia

B.M. ISHMURATOV, L.A. BEZRUKOV, A.F. NIKOLSKY AND I.L. SAVELYEVA

Siberia was joined to the Grand Duchy of Moscow (Moscow Principality, Muscovy, or Russian Tsardom) in a rush for gold and furs. Yet, it never became a new "El Dorado", whilst the all-Russia might was increasingly gaining in momentum. Siberia was destined to become the very first link in the creation of a mighty nation: it formed part of Russia even earlier than the Ukraine or Belorussia did, long before the annexation of the Don river and Kuban river regions, let alone Kazakhstan, the Caucasus, and Central Asia. Siberia never turned into a 'container' for the majority of Russians, though, remaining a sparsely populated land, of little avail for a carefree existence, a "space of opportunities", and a store of resources [1]. Siberia is specializing more and more in the extraction and preprocessing of energy and other natural resources, and in the development of the infrastructure for exporting them to both European Russia and Asian-Pacific countries.

Some figures characterizing the existing distribution patterns of demographic, natural-resource and economic potential across Russia's territory are shown in the table [2].

There are two alternatives of the future long-term investment policies in Russia. The first was implemented during Soviet time and was able to survive in conditions of uncontrolled market relations. Therefore the population distribution in Russia was not to drastically change, and the overwhelming majority of the population was to remain in the European part. Major investments in agriculture, services sector, and in the most highly profitable sectors of processing and engineering industries were also to be concentrated in Russia's European territory. The economy facing the too low prices of raw materials and energy (not counting rent) is squandering, as before, the resources and energy. The regions of Siberia and the Far East are characterized by the top-priority development of industries focusing on the extraction, preprocessing and export of highly marketable raw materials, by a low population densities, poorly developed agriculture, the tertiary sector, food-processing, and light and engineering industries and hence by living standards.

Table

Characteristics of Russia's macro-regions

	Russia total	European Russia	West Siberia	East Siberia	Far East
Population as of 1.01.2001: thou., %	144,339 100	116,617 81	11,753 8	8,922 6	7,107 5
Industrial natural resources: points, %	181 100	62 34	19 11	60 33	40 22
Agricultural area in 1989: mil ha, %, ha per capita	215.7 100 1.5	150.6 69 1.3	35.3 16 2.4	23.0 11 2.5	6.8 3 0.9
Fixed production assets*: mil rubles, %, thou. rubles per capita	5,107,014 100 35.4	4,157,911 81 35.6	315,215 6 26.8	365,527 7 41.0	268,401 5 37.8
Primary agricultural fixed assets*: mil rubles, %, ruble/ha	1,175,124 100 5,448	967,686 82 6,426	113,492 10 3,215	49,791 4 2,164	44,135 4 6,490

*full accounting value, end 2001.
Source: [2].

The other possibility involves attracting the population from European Russia, including migrants from the CIS countries, to the southern regions of Siberia and the Far East. The prices of operating supplies exported from Siberia and the Far East would contain rent; the industrial demand for resources and their consumption would dwindle in the European part. Extra profits (rent) gained from the production of raw materials and sales would be channeled by the regions of Siberia and the Far East towards implementing regional programs aimed at attracting the population and developing agriculture, engineering and other processing industries. Overall, this would also be beneficial for modernization and further development of traditional raw-material-based industriesl.

The latter alternative appears rather unlikely, given the existing situation, the interests of the European center, and Russia's present market conditions. In fact, it would meet best the long-term interests of Russia as a whole. For one thing, further concentration of the bulk of population in the territory of the European part is undesirable both in terms of political and social stability in the region, and due to inevitable acute employment problems there. Besides, one should not forget the strategic disparity between the sparsely-populated eastern Russia and the densely-populated neighboring Asian countries. Secondly, after the collapse of the USSR, 60 % of its agriculturally valuable lands fell outside Russia, with the lands in Siberia and the Far East now accounting for 31 % of all Russian lands. Moreover, current global climate changes on the globe can alter dramatically the stereotypes as to whether western or eastern Russia is best suited more favorable for life and farming. Thirdly, the industrial raw-materials-based status of Russia internationally could hardly be bettered until the raw materials-based status of the Siberian and Far-Eastern regions is beneficial to the development of the various end-product-oriented industries.

Therefore, the Russia-wide geopolitical interests call for a radical shift of its attitude toward the eastern territories. They should no longer be regarded as mere auxiliary sources of raw materials, but as the critical locations for the development itself. This will necessitate attracting the extra population from the western part and employing it to produce the missing industrial and agricultural end products. In the process, Siberia is ready to provide lands for this purpose, as well as natural resources, the poorly utilized funding potential, raw materials, building capacities, uncompleted construction projects, etc. This program promises to become the most important nation-wide program in order to get over the crisis and better the life in Russia.

Siberia has an unprecedented mineral wealth. In the early 1990s, its share in Russia's reserves constituted at least 80 % as regards such resources as coal, natural gas, a number of rare and non-ferrous metals, 70 to 80 % of oil, nickel, muscovite, magnesites, 60-70 % of copper, zinc, molybdenum and cobalt, and over 40 % of gold. In the past, most dramatic examples of the creation of major mining centers of were Kuzbass, the Norilsk region, and the West-Siberian oil and gas complex. This is not impossible at the beginning of the 21st c. Projects on development of the Kansk-Achinsk fuel-and-energy complex situated in the middle of Siberia, on tapping of the mineral deposits in the northern Chita region still are to be implemented, as well as a major promising project titled "Oil and Gas of East Siberia" whose implementation was started by development of the Kovykta gas-condensate field near the Upper Lena. Of vital importance still is the task of providing for a comprehensive exploitation of mining resources, concurrent with controlling environmental pollution [3].

A search of an economic development strategy for Siberia should also be based on analyzing its economic-geographic situation. This implies primarily Siberia's intra-continental macro-position, i.e. its location deep within the Eurasian continent, extremely remote from non-freezing seas and oceans, the country's center, and major internal and external markets to sell its products. Hasty implementation of the market principles in the national economy, its efficient integration into the system of international relations, and the disproportionately escalating transport tariffs in the 1990s were the objective reasons for the fact that the Siberian regions were in the least favorable transport-economic conditions when compared to the other subjects of the Russian Federation.

Because of the obvious advantages of the relatively cheap sea transportation over the more expensive land transportation, the inland and coastal countries and areas enjoy different economic capabilities [4]. This is clearly evidenced by the much higher expenses incurred by transportation of identical products and, hence, extra-priced products and lower price-competitiveness in inland countries and areas as contrasted with the coastal countries. The necessity of continually spending its surplus product in order to compensate for higher transportation costs and to adjust to them is the reason why most economic difficulties are faced by inland Russia and, first of all, by "super-continental" Siberia, most of Russia's inland expanses.

Indeed, looking at Russia's remoteness from oceans, seas and year-round navigation sea-ports, one finds that virtually the whole of Siberia is at a distance of more than 2 thou. km from them, i.e. is located in the ultra-continental zone. Taking into consideration the particular distances by railroad lines, however, the remoteness of Siberia's "core" between Omsk and Chita from major Russian year-round sea-ports (St. Petersburg, Novorossiisk, Murmansk, Nakhodka, etc.) will amount to over three thou. km, while Krasnoyarsk Territory and the Irkutsk region are at a distance of more than four thou. km.

It is important to stress at this point that at such a large distance from seas and oceans not a single major country in the world, except for Russia, includes regions that have achieved a level of industrialization and urbanization which is typical of the most advanced Siberian regions and areas. Given the seemingly sparsely populated area and a poor degree of development, the share of Siberia in Russia's main economic indices is rather impressive: 22.7% in the gross regional product as of 2000; 26.2% in the volume of industrial production; 32.2% in railway cargo transportation; and 39.1% in the output of raw materials. On the other hand, in China the extensive Sinjang and, moreover, Tibet that are at more than 2 thou km from the sea coasts, are distinguished by a very low industrial potential. The territory of Canada, at identical distances from the unfreezing sea ports (the coast of the Arctic Ocean, and some of its islands) has an extremely low degree of development and settlement. The USA, Brazil and Australia that are at the top in the list of the states in their area, have no such inland spaces at distances more than 2 thou km from the sea (the most inland regions of the two last countries are also characterized by a low level of economic development and sparse population).

What are the methods, procedures and techniques used to organize the economic activities which enabled Siberia to make appreciable progress when compared with the other inland macro-regions of the world? Based on the need to drastically reduce the high transport costs, it would be logical to build the economy of Siberia through the use the following adaptive principles of an economic-geographic nature: the organization of the economy on a well-thought district basis and complex-formation, a firm establishment of the territorial division of labor and the attainment of a maximum possible level of self-sufficiency of the Siberian economic regions (and sub-regions), "adjustment" (attraction) of the population and economy to the critical traffic routes, realization of, primarily, the nearest intra-economic and, partly, inter-regional relations in respect of the exchange of products, with a certain limitation of the less efficient distant and ultra-distant relations, etc. It should also be noted that some of the above principles were used (though not always) by economic practices, both in a planned (target-oriented) fashion and spontaneously [5].

The largely successful effort to organize Siberia's economy (as is the case, by the way, with all Russia – USSR) based on the regional requirements and the scientific principles of economic regionalization can serve as a most illustrative example of a large-scale application of the planned adaptive methods to the development of continental-type areas. According to the Soviet economic regionalization theory, the optimization of the intra-regional division of labor leads to an increasingly higher 'complexity' of the economy and self-sufficiency in every large region, as well as restraining the remote international and inter-regional relations and exchanges. Because of the shortened mean distances of transportation, this would reduce transportation costs thus mediating the negative effect of the factor of land distances.

In the Soviet era, the economic regionalization, in the case of Siberia, proved to be a major tool for resolving challenging economic problems and implementing formidable projects, including the erection of large factories, based on which, various energy-production cycles (EPC) were developed and large-scale territorial-production complexes (TPC) created. Suffice it to mention the existence such an inter-regional complex as the Ural-Kuznetsk integrated plant, the shaping of the West-Siberian oil and gas complex, the emergence of a whole system of Angara-Yenisei TPCs: the Irkutsk – Cheremkhovo, Bratsk – Ust-Ilimsk, Central-Krasnoyarsk, and Sayansk complexes.

Another important adaptive method of development of Siberia's vast expanses that has in many respects established spontaneously, is the technique of clearly pronounced linear (band-like or "corridor"-like) development. The effect of this method manifests itself in an enhanced role of linearly ordered economic-settlement structures. They are produced almost exclusively on the basis of the major economic axes: main transportation routes with fixed direct-influence zones, where the situation involves gaining higher economical and social effects than is the case with adjacent territories. This effect is built upon the immense economic advantages of "magistralization" and specialization of the transport routes and communication, because transportation by mainlines is several times and, perhaps, even several tens of times, as cheap as that by the rest of the network. Consequently, enterprises and human settlements situated along trunk routes and in zones of their direct influence (i.e. along economic axes) incur the least transport expenses and have the best opportunities as regards their further development and growth.

Linear-territorial systems of productive forces are being shaped in Siberia, based chiefly on such primary economic axes as railway mainlines and navigable rivers (water ways). By the early 20th century, they concentrated the unquestionably overwhelming portion of Siberia's city-type settlements (500 of 575) and their population (nearly 97 % of the 17.8 mil city inhabitants). Especially high is the urban-forming influence of the railway mainlines, along which a total of about 90 % of the macro-region's city population lives to date. Thus, the relatively narrow 40-kilometer-wide strip along the Trans-Siberian Railroad stretching 5 thou. km from the western boundary of the Tyumen region to the eastern boundary of the Chita region and accounting only for 2 % of Siberia's territory concentrates 43 % of its city population.

Somewhat less important, in terms of the urban-forming properties, when compared with the railroads, yet a still important role is played by the water ways, accounting for approximately 80 % of Siberia's city inhabitants (a considerable number of cities and towns are located along both the railroads and water ways, belonging in the two categories). Along the banks of only five large rivers: the Ob, Yenisei, Angara, Tom, and Irtysh, over 55% of the urban inhabitants are concentrated along the banks of only five large rivers: the Ob, Yenisei, Angara, Tom, and Irtysh. With the inclusion in this list of the Tura, Selenga, Lena, Ingoda, and Chulym, the share of the macro-region's urban population living along the ten aforementioned rivers, reaches 2/3.

On the other hand, it can be stated that although Siberia in Soviet times had a generally adapted (to the continentality conditions) territorial organization of the economy and population, a great deal remains to be done in order to eliminate serious deficiencies and disproportions. Since the patterns and principles of economic regionalization in the USSR used to be largely grossly violated, master economic regions never received any comprehensive, self-sufficient development, the shaping of the Siberian TPCs was often dominated by a sectoral (departmental) approach, their EPCs usually lacked any efficient completion in the form of upper "stories" of production facilities, etc. The Siberian regions were mainly oriented toward exports of raw materials and semi-finished products to the European part of the country and, partly, to other countries, satisfying through imports their own requirements for various end products. At the same time, the commodity exchange between the Siberian regions themselves (including adjacent ones) was poorly developed, and the share of long-distance and even very long-distance transportation was increased, which was obviously in contradiction with the adaptive principles of organization of the economy in the inland regions.

The existing serious deficiencies as regards Siberia's economic organization in the Soviet period still were partly offset by the mechanism of planned centralized redistribution of the surplus product created by the national economy, including state regulation of pricing in the basic industries, primarily in th sphere of power generation and transport. In the post-Soviet period, however, with the dramatic reduction of the controlling and redistribution functions of the state and Russia's impetuous integration into the world market, the adverse influence of Siberia's inland position on the effectiveness of its economy has dramatically reduced.

As a result, the share of transportation costs in the final value of Siberia's products reaches an unprecedented amount for the overwhelming majority of Russia's regions, not to mention the other countries of the world: up to 50 % of the value of many kinds of large-size products, and up to 70 or higher for some of its kinds. The exorbitant transportation constituent of exported raw material and semi-product costs significantly reduces the profitability, earning power and price competitiveness of products for most basic industries and sectors of the Siberian economy. Because of the special significance of the export-import and, partly, inter-regional economic relations, a very sizable part of the financial resources is "evacuated" through payment of transport expenses from the income of producers. Since the high transport outlays ultimately lead to a marked decrease in taxes paid to regional and local budgets, the Siberian subjects of the Russian Federation can be regarded as the most "frustrated" regions of the country, as regards the unfavorable influence of the transport-economic factor, on the size and balance of their budget revenues and expenditures.

Thus the objectively raised transportation costs in regard to products are a sort of negative differential rent of the location which is continuously escaping from Siberia at its interaction with Russia-wide and, especially, international market. With the other conditions unchanged, the surplus product lost by the Siberian economy, which is equal in value to the negative rent of the location, will be the greater, the more tightly the economy of this "ultra-continental" macro-region is involved in the system of international "oceanic" exchange.

While in the past Siberia's internal potential was a chief actor of its development, at present (with the increasingly extensive inclusion in the world market) the macro-region's economic capabilities and opportunities are to a still greater extent determined (or, more exactly, limited) by its economic-geographic position on the glob, and by its place in the hierarchical structure of a polarized mutually linked world.

It is therefore strategically important to look for such a more efficient system of Siberia's economic ties and exchanges where the share of raw materials and semi-finished products in its export would be consistently decreasing, with a concurrent increase of the share of more expensive, transportable finished products. A central direction of Siberia's further economic development and improvement of its effectiveness, with due regard for its inland macro-location, should be a deepening of a processing of raw materials by setting up final redivisions, primarily using the existing industrial base. A large positive role in Siberia's economic "rapprochement" to the leading centers of the country and the world, and to the seas and oceans would be played by the creation of trans-continental transport corridors along the Trans-Siberian Railroads and BAM as a means of a further magistralization of the railroad transport and a drastic reduction of transportation costs.

REFERENCES

1. Ishmuratov B.M. Siberia in the Russian and World Perspective (Essays on Socio-Economic and Political Geography). – Irkutsk: Ottisk publ. 2003, 169 p.
2. Nikolsky A.F. The Geoeconomic Reproductive Process: The Theoretical Foundations and Management Principles. Irkutsk: SB RAS IG Publisher, 2004, 165 p.
3. Savelyeva I.L. Mineral Resource Production Cycles: Problems of Regionalisation and Rational Nature Management. Novosibirsk: Nauka, 1988, 133 p.
4. Bezrukov L.A. The transport-geographical continnetality of Russia: the concept, assessment, dynamics. IZV RAN. Ser. Geogr., 2004, No. 5, pp.15-20.
5. Bezrukov L.A. The transport-geographical continentality of Russia: The transport-intensity of the economy, and adaptive consequences. Izv. RAN. Ser. Geogr., 2005, No. 3, pp.48-55.

CHAPTER

9

History, Characteristics and Objectives of Geographical Resources Science in Russia

L.M. KORYTNY AND YU.P. MIKHAILOV

BASIC CONCEPTS

If the object of geography is taken to be the natural and social-economic (humanitarian) life environment of society and its subject is assumed to be the explanation of how its territorial organization has emerged and evolved, under the influence of what factors it changes from place to place, and how these differences show up in the life of society, this inevitably brings up the question: What are the individual factors, prerequisites and conditions of this environment that can be regarded as having the role of the cause of these changes? Among the most important of them are the natural resources. In a more comprehensive sense, resources signify the means, possibilities and sources of the goods and utilities that serve to meet some demands of human beings in their life activities.

Natural resources have always been a key object for study of economic geography. The spectrum of natural resources is very broad, and this makes itself evident not only in the diversity of natural resources as regards their material (substance) composition but also in the time, place and magnitude of demand for them. To put it otherwise, assigning a particular natural phenomenon to the category of natural resources depends not only on its belonging to naturally occurring objects but also on economic, social and other subjective factors and causes, and this adds a great deal of complexity to the assessment of resources by introducing uncertainty into it. Given the existing definitions of natural resources, this factor is ignored, whereas it raises in quite a new fashion the issue of inventory, and qualitative and quantitative, and, primarily, economic assessment of natural resources. For Siberia, this issue is an especially urgent problem, as a number of its natural wealth, for a great variety of reasons, cannot still be involved in utilization and, as yet, can be regarded as nothing more than a free good. One further condition also deserves mention, which precludes assigning a particular good to the natural resources: it implies the unlimited reserve of the good where any one of the units of this reserve is of zero value for a given period of time if one proceeds from the marginal utility theory.

There is a need for a clear awareness of the fact that the natural resources as they are naturally occurring phenomena are an economic category as well. Without regard for this, one cannot form a correct idea of their role in the life and evolution of society, its resource potential, and of the rational use of this potential when applied to resolving the optimization problem of the territorial organization of the life of society, and this is the major objective of georesources science.

We believe that georesources science combines the seemingly different natural and socio-economic components pertaining to different branches of geography and, on the other hand, forming an inseparable unity. Such a vision of georesources science seems highly important to us, as it clearly indicates its position in the system of geographical sciences. Ignoring this unity, as well as the differences, would inevitably lead to crude constructions and conclusions as regards the assessment of natural resources which is not infrequently hypertrophied. Hence we have false theoretical constructions as well as misleading conclusions and recommendations for the utilization of natural resources. Furthermore, it must be emphasized that we by no means want to belittle the importance of the natural resources in the life of society – it is, of course, large. But returning to the times when a major measure of development of the society and state was an ever increasing tapping of the sources of natural resources would undoubtedly be a step backward.

Generally speaking, as rightly pointed out by P.Ya. Baklanov [1], "of fundamental importance is the fact that ... numerical estimates and criteria cannot be regarded as determining" [p.10].

CLASSIFICATION OF NATURAL RESOURCES

Any assessment of the role played by natural resources in the life of society, as the study of any other complex phenomenon, cannot obviate the need for their classification and typology. The spectrum of natural resources is very broad and is continually incorporating new, quite diverse, elements, and this renders the problem of classification of the resources quite difficult. Therefore, it is entirely obvious that there can be several classifications built upon different bases according to the objectives, for achieving which a particular classification is created.

It is known that all natural resources are subdivided into two large groups: reproducible and non-reproducible. Assigning any resource to a particular group according to this attribute would not be a matter of concern, but this is important, because in the process of utilizing these resources the attitude to them can alter drastically.

Another system of classification is built according to the way in which the resources are consumed. Some natural goods are consumed by man directly from the environment (air, water from natural sources, sunlight and heat, etc.). Following D.L. Armand [2], they can be called the resources of environment. The other resources, before they are consumed, must be withdrawn from Nature as objects of labor and must let pass through the stage involving their conversion to raw materials (all energy carriers, metal ores, many agricultural products, forest, etc.). Finally, it is possible to distinguish the natural resources by the spheres of their predominant usage, such as in the agrarian sector, in industry, in the sphere of social welfare and consumer services, etc. It is pertinent to note that such a subdivision is relative in character, as one and the same resource may well fall within several classification groups.

Natural resources are of paramount importance in the life of society; we believe, contrary to a widespread opinion, however, that they do constitute a prerequisite, important as it is, rather than a factor of societal development. Whilst there are states and regions prospering by selling out their natural resources, such as some of the Middle-eastern countries, this is merely an exception to the rule, and such prosperity is insecure. Much more representative is another regularity. Natural resources endowment can have an adverse influence, paradoxical as this may be, on the social-and-economic development of society. It is for that reason that the term "the resource damnation" was coined, which is referred to the states that squander their national wealth. Among them is also the present-day Russia whose budget depends critically on exporting petroleum, gas and timber, and on sales of fish catch quotas, and other crudes. This resource component is most conspicuous in Siberia and in the Far East. On the other hand, there are states on the globe which virtually lack any significant natural resources, such as Japan, The Netherlands, Belgium, Danemark, Island, and others, and are forced to purchase many natural resources from abroad, but are highly developed countries.

To gain a correct understanding of the place occupied by the natural resources in the life of states and regions is the key objective of georesources science. The resolution of this problem is important for Siberia and the Far East which, to a greater extent than the other regions of the country, need a further enhancement and widening of the scope of research into the natural environment and its resources.

THE ADVANCEMENT OF GEOGRAPHICAL RESOURCES SCIENCE IN RUSSIA

Natural resources in all countries have always been utilized and explored; in Russia, however, a special role was played by the resources of its eastern regions. Interest in their natural resources was aroused as early as before the rule of Peter the Great. M.V. Lomonosov's aphorism that the Russia's power would accrue through Siberia is a matter of common knowledge – a statement full of hopes and optimism. Yet it was not until the second half of the 19-th century that the Siberian natural environment and its wealth became the subject of extensive, purpose-oriented study that was elicited by the emergence and advancement of large-scale industry, transport, trade, the Russia-wide market, and Capitalist relations in the country in general. A significant contribution to the exploration of Siberia was made by the Russian Imperial Academy of Sciences, the Russian Imperial Geographical Society, the Re-settlement Committee, various private companies and entrepreneurs, as well as by the works of such outstanding scientists as D.I. Mendeleev, P.P. Semenov-Tian-Shansky, V.I. Vernadsky, V.A. Obruchev, and others.

Large-scale research into the natural environment and its wealth in Siberia and the Far East traces back to the end of the 1920s and to the 1930s. It was pursued by A.E. Fersman, S.V. Obruchev, A.I. Gubkin, V.L. Komarov, and other eminent scientists. Among the works summarizing the achievements from those years, "The natural-historical regionalization of the USSR" by S.G. Strumilin and I.S. Lupinovich [3] deserves mention. A great deal in this area was done by N.N. Baransky and his disciples. In the late 1940s, N.N. Kolosovsky offered his concept of energy-production cycles [4]. A significant importance in the concept was attached to the linkage between the economy and natural resources. It should be remembered in this case that the concepts of "territorial production complexes", "energy-productive complexes" and "energy-production cycles" are built largely upon singling out "... the typical stable sets of production processes that arise in a mutually conditioned (co-subordinate) fashion around the main process of a given form of energy and raw materials" [p. 42], i.e. they were based on the feasibility and, more likely, even technological principle. Therefore, these concepts can be truly referred to geographical resources science with some reservation only, although it goes without question that they contributed to the advancement of economic geography and georesources science in particular. The cited references were all characterized by a search for new approaches to pursuing georesources science as a new independent branch of geography lying at the junction of natural and human geography.

Yu.G. Saushkin [5] was the first to point out the need to create georesources science and to give its definition. He defined it as "... a special geographical discipline dealing with the study and allocation of territorial complexes of natural resources and addressing the issues related to their utilization. This branch of science lies at the interface between physical geography and other natural sciences, on the one hand, and economic geography, on the other" (p. 430). Clearly, this interpretation is insufficient from the present-day standpoint, correct as it is. In our opinion, the publications dealing with georesources science fail to point to the important role played by the socio-economic environment in the assessment of natural resources.

In the late 1960s, geographical resources science was further pursued at the IG USSR AS under the guidance of A.A. Mints. In his numerous papers and fundamental monograph [6], he was the first to summarize – from the geographical standpoint – the experience gained from studying and assessing the natural resources and pioneered the formation of the conceptual-terminological apparatus for georesources science. He brought into general use such notions as "natural resources", "qualitative, quantitative and economic assessment", "territorial combinations of natural resources", etc. Together with his disciples, he made an attempt to assess the country's natural-resources potential and singled out territorial complexes of natural resources [7]. Noteworthy is also the work done by I.V. Komar [8], who introduced into the conceptual-terminological apparatus the notion of natural-resources cycles thus extending N.N. Kolosovsky's concept.

It might be as well to point out that the issue of economic assessment of natural resources remains debatable to the present day. The method of normative assessments, in wide use in the past, provoked abundant justified objections, and in the new market and private property circumstances is impractical altogether.

Georesources science was also pursued at the Academy's other research centers. Thus, at the Pacific Institute of Geography of the Far-Eastern Division (FED) of the USSR Academy of Sciences, in the late 1970s and during the 1980s P.Ya. Baklanov [9] developed the concept of territorial natural-resources systems, with special emphasis on the analysis of territorial-aquatic combinations of natural resources. Economic assessments of the Far-Eastern resources were then incorporated taking into account not only the cost components but also the value components related to consumer qualities and competitiveness of the resource. A particular significance was attached to a consideration of (along with direct resources consumption) reverse resources consumption by which it was meant a change in natural resources, with the discharge of production waste. Far-Eastern scientists furnished specific assessments of the mineral-raw material, forest, land, shelf and water resources.

In the same years, at the Institute of Geography (IG) SD USSR AS, the laboratory for assessment of natural resources was specially set up under the guidance of Yu.P. Mikhailov, with the objective to develop the theory and practice of geographical resources science. A major effort made by the head of the laboratory and its staff members (S.T. Budkov, A.I. Aseyeva, V.M. Parfenov, V.L. Yulinov, A.R. Batuyev, V.V. Rudsky, V.A. Chervyakov, and others) went into establishing a dependence of the natural potential of lands and other resources on the geographical location, landscape structure, ecological limitations of utilization, and on other factors. The investigations were carried out in the Sosvinskoye Ob region, in the valley of the Lower Irtysh, in the Irkutsk-Lena region, and in other areas. A special research area, bioresources science, was pursued by L.N. Ilyina, who pioneered the analysis of the relation between geography and resources science [10]. An important problem of the use of solid waste as resources is pursued by T.I. Zabortseva [11]. Particular emphasis in the investigations carried out in the division is placed on the topics related to mapping natural resources. A classification of resources maps was, for the first time, suggested [12]. The field concept-based technique of contour mapping due to V.A. Chervyakov was accepted throughout the country [13]; numerous maps of the Asian Russia and its regions were generated.

The Council for comprehensive study of taiga territories was set up at the IG SD USSR AS in 1963. The Council held enlarged meetings, the materials from which were published in 18 collections. In them, issues related to the study of natural resources received primary emphasis [14]. Important proof of the urgency and success of the georesources studies at the IG SD USSR AS was the creation in 1980 of the journal titled "Geography and Natural Resources".

The 1980s saw the erosion of interest in topics of georesources science in the USSR. That was manifested primarily as a drastic decrease of the number of publications, and theoretical, summarizing and comprehensive books. The sole exception was the monograph by the Ukrainian authors N.G. Ignatenko and V.P. Rudenko [15] on integral assessment of the natural resources potential. Whilst investigations into separate kinds of resources from the geographical standpoint were, of course, continued. First and foremost, mention may be made of a number of investigations that were carried out at the IG SD USSR AS under the guidance of B.M. Ishmuratov into energy-production and resources cycles: mineral-raw material (I.L. Savelieva), forest-energy (E.V. Latina and S.N. Sokolov), water system management (L.A. Bezrukov), etc.

PRESENT STAGE OF GEORESOURCES SCIENCE IN RUSSIA

At the present time the natural environment – man interaction on our planet earth is characterized by a special complexity and large-scale and dynamic character. Although Man since antiquity had satisfied his needs by transforming his outside ambient via extraction of natural resources therefrom, it was not until the last several decades that the traces of this transformation became manifest for all levels and regions. Civilization is getting involved in an increasingly deeper global crisis that shows up as the resources "starvation", the population explosion, progressively exacerbating pollution of natural environments, a disturbance of biochemical cycles, a decline in the biodiversity, etc.

The crisis in Russia in recent years has its characteristic properties. Wasteful consumption of all kinds of resources that came to be a major cause for economic failures, for the restraint of the scientific and technical progress, the low life quality of most of the population, and the impairment of the ecological situation in Soviet times, actually still persists in the new Russia. The country lives mainly by "squandering" the natural resources of its citizens – both living nowadays and future generations. Concern for environmental protection and public health, and for rational use of natural resources is constantly put on the back burner, and the situation is further aggravated by the obviously unsuccessful reorganization of the environment management system in 2000.

One of the reasons for such a situation has to do with the methodological concept in addressing the issues of resources utilization that is inadequately adapted to the contemporaneous complicated socio-economic conditions. It is the development of a new concept that should be the chief goal of georesources science. In our opinion, this development in regard to natural resources was already initiated in the 1990s in the following areas.

1. **A new approach to the essence of natural resources.** Essentially, it signifies the awareness and recognition of the fact that natural resources should include all components and properties of natural geosystems, once they are treated within the aspect of their potential use for any (not only productive) man's activity. The notion "environmental resources" is introduced into economic usage as "a totality of environment-forming components that ensures an ecological equilibrium in the biosphere, and a normal living environment for man" [16, p.59], since environment-forming functions are items of real value. The source of such an approach to the equalization of material and non-material requirements can be found in the works by the physiographers D.L. Armand [2], V.S. Preobrazhensky, Yu.K. Yefremov, and others, in publications of L.N. Ilyina [10], and in the new definitions of the notion "natural resources" by V.M. Razumovsky [17], N.F. Reimers [18], and L.M. Korytny [19]. This research area is gaining an increasingly greater prevalence also in the economy of nature management; specifically, it forms the basis for ecological economics and is closely associated with the theory of natural capital; the assimilation potential of natural environment, the carbon credit, etc. are treated as special kinds of natural-resources potential.
2. **Treating the territory and water surface as a resource.** Land and water surface areas are known to be traditionally involved in the notions of "land resources" (for agricultural nature management, for example), and "water resources" (for water transport, for example). On the other hand, the territory and water surface are the reservoir for all natural resources, the operational basis for any economic and other activity of society, and as such, they can be treated as a special territorial-water surface resource [20-22]. This is of particular current importance in today's Russia when the terrestrial and water surface are only becoming a valuable commodity, an item of political and economic bargaining. The regi9onalization processes of the economy and management are being enhanced. The economic-geographical,

primarily transport-geographical, location that is also a distinctive resource, is becoming a critical economic factor.

3. **Natural-resources regionalization.** The ideas of such a regionalization were theoretically substantiated in publications of A.E. Fersman, S.G. Strumilin, A.A. Mints, Yu.D. Dmitrevsky, and others. During the last decade they have been successfully pursued by I.L. Savelieva [23], treating the essence of the regionalization in the context of paragenetic interrelationship and mutual conditionality of different-quality natural resources where at different levels the leading criteria for identifying the regions can be different. Within Russia, she identified 31 fractional integral districts, with the geological-structural composition being taken as the leading criterion. P.Ya. Baklanov [24] backs up the concept of different-level regionalization and applies it for North-Eastern Asia; he identifies both large areas within sea and major river basins and medium-size areas – in the form of portions of larger areas, as well as territorial, aquatic-territorial and aquatic systems – as potions of medium-size areas. An important point is that at the early stage it is recommended that all areas should be singled out without regard for state borders. The basin principle of regionalization, including the natural resources regionalization, is substantiated by these authors in [25] primarily through delineation – within a basin – of resource flows, and by the natural character of the boundaries, or watersheds.
4. **Integral assessment of the natural resources potential.** Significant difficulties associated with quantitative assessment – both natural and economic – continue to escalate with every attempt at integral assessments of the potential to such an extent that this has long become a highly unpopular matter. Nevertheless, such efforts are called for, because the best avenues for the exploitation of the potential could be chosen and investors in doing so attracted only through a thorough understanding of all resources of the territory by comparing their significance to the structure of the integral potential, by assigning priorities and identifying territorial combinations of the resources, and with due regard for environmental factors and limitations. Such an assessment within the administrative and intra-regional economic areas was made for the first time at the Institute of Geography SB RAS for the Irkutsk region, based on the final result-approach, i.e. by taking into account the prices for marketed commercial products [26]. Based on this technique, a start has been made on integral assessment of the potential in other subjects in the east of Russia as well.
5. **Territorial and historical analysis of resource utilization.** In the second half of the 1990s, georesources science research was resumed by scientists from the IG RAS. First of all, mention should be made

here of a series of papers authored by G.A. Privalovskaya [27, 28, and others], analyzing the role of resources utilization in connection of a globalization of the economy and with the need to switch over to sustainable development. For Russia, it has been clearly demonstrated that the natural resources have been and are the major accelerator of a growth of the economy both at present and for a long period in the future; furthermore, the challenge of paramount importance facing georesources scientists has been and is to study the regional differentiation of the dynamics of resources utilization, concurrent with consideration for its impact on environment. A substantiation was provided for the Russia-specific resources space in which a special role is played by Siberia. D.I. Lyuri [29], extending the ideas of resources cycles suggested by I.V. Komar, was able to relate the development patterns of resource utilization in the historical perspective to the factors giving rise to ecological crises; it was well documented that regional crises would not always end in disasters.

6. **Mapping of natural resources.** Research of this kind has not discontinued throughout the last several decades. Many years of collaboration among physiogeographers, economic geographers and cartographers of the IG RAS and from other institutions resulted in two fundamental atlases: the atlas of the world's snow-ice resources, and the atlas of the world's natural resources [30, 31]. Of course, they constitute a huge contribution made by Russian scientists to international resources science, demonstrating their leading role in this area of geographical science. In recent years Moscow State University, IG RAS and other organizations have successfully pursued electronic resources mapping, primarily in addressing issues related to management of a balanced socio-economic development [32, 33].

Thus the analysis made here has shown that Russian resources science is currently on the rise because of the urgency of challenges facing geography as regards the study of formation patterns of spatial structures of environment-society interaction, the role of resources utilization in this process, as well as the huge theoretical and procedural potential accumulated in this area of geography. That was confirmed by the international conferences held at Irkutsk within the framework of the Second and Third Baikal Economic Forums that were consistently devoted to a further development of the methodological and applied aspects of the study on the resource management [34, 35].

The immediate goal of resource management is to help reform resource utilization so that it would not only yield an adequate profit to the State by exporting its products but also actively contribute to resolving nationwide strategic tasks, primarily implying switchover to the innovation route of development, based on reconciling the nationwide, corporate and regional interests. In this case the chief objective of resource management should include:

a) to improve regionalization as the only method to account for the spatial resource diversity in order to decide on the territorially differentiated strategy and tactics of resource management;
b) to develop the technique of total calculation of the natural rent, with due regard for the rent of the geographical (including economic-geographical and geopolitical) location;
c) to continue integral assessments of the natural-resource potential of the regions, with the obligatory inclusion of the ecological potential in it; and
d) to create maps and atlases of the state, assessment and utilization of resources of any kind, within the content of their dynamics, based on state-of-the-art GIS technologies.

REFERENCES

1. Baklanov P.Ya. The Far-Eastern Region of Russia: Problems and Prerequisites of Sustainable Development. Vladivostok: Dalnauka, 2001.
2. Armand D.L. To Ourselves and Our Grandchildren. Moscow: Mysl, 1966.
3. Strumilin S.G. and Lupinovich I.S. Natural-Historical Regionalization of the USSR. Moscow, 1947.
4. Kolosovsky N.N. The production-territorial combination (complex) in Soviet economic geography. Vopr. geografii, 1947, No. 6.
5. Saushkin Yu.G. An Introduction to Economic Geography. Moscow, 1958.
6. Mints A.A. Economic Assessment of Natural Resources. Moscow: Mysl, 1972.
7. Mints A.A. and Kakhanovskaya T.G. Experience of a comprehensive assessment of the natural-resources potential of the regions of the USSR. Izv. AN SSSR. Ser. geogr., 1973, No. 5.
8. Komar I.V. Rational Utilization of Natural Resources, and Resource Cycles. Moscow: Nauka, 1975.
9. Baklanov P.Ya. Direct and reverse Natural-Resources Links in the Structure of Territorial Industry Systems of the Far East. The Economic-Geographical Aspect of Nature Management in the Far East (Questions of Methodology and Techniques of Resources Assessment). Vladivostok, 1979.
10. Ilyina L.N. Geography and resources science. In: Methodological Questions of Geography. Irkutsk: IGS & FE SD USSR AS, 1977.
11. Zabortseva T.I. The modern structure of procurement and processing of secondary material resources in regions of the Near North of East Siberia. In: The Natural-Resources Potential, and Nature management. Irkutsk, 1989.
12. Mikhailov Yu.P. Territorial resources as a geographical notion. In: The Natural-Resources Potential of the Asian Russia and Bordering Countries: Avenues for the improvement of utilization. Irkutsk, 2002.
13. Chervyakov V.A. The Field Concept in Modern geography. Novosibirsk: Nauka, 1978.
14. Mikhailov Yu.P. Present and future problems of studying the taiga. In: Siberian Geographical Collection. Moscow-Leningrad: Nauka, 1965.
15. Ignatenko N.G. and Rudenko V.P. The Natural-Resources Potential: Geographical Analysis and Synthesis. Lvov, 1986.
16. Lukyanchikov N.N. and Potravny I.M. Economy and Organization of Nature Management. Moscow: Troika, 2000.

17. Razumovsky V.M. Ecological-Economic Regionalization (Theoretical Aspects). Leningrad: Nauka, 1989.
18. Reimers N.F. Nature Management. Glossary. Moscow: Mysl, 1990.
19. Korytny L.M. Natural-resources problems and their classification. Geografiya i prirod. resursy, 1992, No. 1.
20. Korytny L.M. Territorial resources: definition, classification, approach to assessment. Geografiya i prirod. resursy, 1995, No. 3.
21. Maksimov G.N. Toward the content of the notion "territorial resources". Geografiya i prirod. resursy, 1997, No. 2.
22. Mikhailov G.N. Territorial resources as the geographical notion. In: The Natural-Resources Potential of the Asian Russia and Bordering Countries: Avenues for the Improvement of Utilization. Irkutsk: Izd-vo IG SB RAS, 2002.
23. Savelieva I.L. Natural-resources regionalization of Russia. Geografiya i prirod. resursy, 1997, No. 1.
24. Baklanov P.Ya. The natural-resources potential of North-Eastern Asia: problems of its assessment and rational utilization. In: The Natural-Resources Potential of the Asian Russia and Bordering Countries: Avenues for the Improvement of Utilization. Irkutsk: Izd-vo IG SB RAS, 2002.
25. Korytny L.M. The Basin Concept in Nature Management. Irkutsk: Izd-vo IG SB RAS, 2001.
26. The Natural-Resources Potential of the Irkutsk Region. Irkutsk: Izd-vo SB RAS, 1998.
27. Privalovskaya G.A. Resources utilization in the Russia's modern conditions. Izv. RAN. Ser. geogr., 1999, No. 3.
28. Privalovskaya G.A. Resources utilization in the Russia's modern economic space. Izv. RAN. Ser. geogr., 2002, No. 2.
29. Lyuri D.I. The Development of Resources Utilization and Ecological Crises. Moscow: Izd-vo Delta, 1997.
30. Atlas of the World's Snow-Ice Resources. Moscow- Kiev, 1998.
31. Nature and the Earth's Resources. Moscow – Vienna: Ed. Helzel, 1998, v. I-II.
32. Korytny L.M. and Batuyev A.R. Mapping of resources factors of regional development. In: Cartographic and Information Support of Regional development Management. Irkutsk: 2002.
33. Atlas "Irkutsk Region: Ecologial Conditions of Development". Moscow-Irkutsk, 2004.
34. The natural-Resources Potential of the Asian Russia and Bordering Countries: Avenues for the Improvement of Utilization. Irkutsk: Izd-vo IG SB RAS, 2002.
35. The Natural-Resource Potential of the Asian Russia and Neighboring Countries: Geoeconomic, Geoecological and Geopolitical Regionalization. Irkutsk: IG SB RAS Publisher, 2004.

CHAPTER

10

Natural Economic Regionalization of The Asian Part of Russia

I.L. Savelyeva

The term "natural economic regionalization" was coined and introduced by Yu.G. Saushkin in 1980. He gave reasons for using it in lieu of the then widely accepted terms "ecological economic system", and "ecological economic district", namely, it more nearly approximated the workings of the nature-society interaction. On the other hand, he pointed out that Russia's regionalization that was carried out as early as 1880 by P.P. Semenov-Tian-Shansky, the regionalization of the USSR, implemented by Gosplan (State Planning Commission) under the guidance of G.M. Krzhizhanovsky, as well as the regionalization undertaken by N.N. Kolosovsky (1947) were essentially natural economic in character, based on taking proper account of the characteristics of natural environment as a basis for development of a specialized (diversified) economy or for the setting up of territorial-production complexes.

To undertake the country's natural economic regionalization Yu.G. Saushkin analyzed 16 factors; of them, nine factors related to natural resources, and seven factor of a socio-economic nature. As regards the natural-resource factors, he attached a significant role to geological (tectonic) structures, some of which have or will become "cores" or "fields" of economic regionalization. He argues that superposition of different factors on one another in a particular proportion lies at the basis of the formation "mechanism" for natural economic districts, and the purpose that is served by regionalization itself is to optimize the linkages within the nature-economy system [1].

Natural economic regionalization is a specific kind of socio-economic regionalization. Depending on the scope of the studies undertaken, it can be either intermediate between natural resources and economics, i.e., according to N.N. Baransky [2], with a focus on accumulating an adequate amount of evidence to be used in integral economic regionalization, or address independent issues related to an improvement of the structural organization of production, and to an optimization of nature management.

We understand by a natural economic district (NED), regardless of its rank, such an area of the terrestrial surface, including the subsurface area, that is relatively uniform, on the one hand, in intrinsic linkages of the different-quality natural conditions and resources (mineral, fuel-and-energy, land, soil-climatic, water, and biotic) and, on the other, provides a sustainable (for given conditions) interplay between the economic activities and the potential process of economic development. It is a definite taxon of territory that has a specific natural and economic content. Furthermore, nature is also regarded as an environment that has been modified through previous human activities, and as the source of resources, the 'bridgehead' (the basis) for a further concerted development of the economic system.

The chief principle of natural economic regionalization implies that each district should be provided with a definite type of economic development of the territory according to environmental characteristics, the resource potential, and the economic-geographical location. Equally important in this regard are also the natural peculiarities, but a detailed treatment of them is addressed by a special branch of natural resource regionalization [3, 4]. In the event of implementing natural economic regionalization, the boundaries of natural resource districts that are set n accordance with the particular landmarks, are updated according to territorial distribution of pre-existing or eventual production systems.

Natural economic regionalization, like natural resource regionalization, can be integral and component-wise in its structure. In the case of component-wise varieties of natural economic regionalization (forest-economic, agro-ecological, and associated with the formative conditions of mineral and raw-material cycles of production, and others), the objective if to address and solve some particular problems of optimizing structural organization of separate sectors of the national economy, and rationalizing the utilization of their basic resources in accordance with the concept of resource cycles suggested by I.V. Komar [5]. Integral natural economic regionalization of the particular territories is to improve structural organization of the entire national economic complex. The concept of energy production cycles suggested by N.N. Kolosovsky [6] fits in well with such a line of inquiry. The concept makes it possible to implement two main avenues for nature management optimization: by minimizing the consumption of resources, the output of waste and by-products and, hence, the adverse environmental effects, and by maximizing the marketable output through territorial cooperation of the various industries.

Natural economic regionalization, built around the concepts of resource and energy production cycles, and based on the theory of formation of territorial-production complexes, allows definite conclusions about the merits and demerits of structural organization of nature management in a regional setting and, in the final analysis, it permits this type of regionalization to be treated not only as an intermediate stage or as one of the basic stages of integral economic regionalization but also as quite an independent investigation into structural organization of economic systems.

The procedure of any regionalization on its own is a unified process of identifying some typical territories and delineating the boundaries between them. In this connection, the principles of integral and component-wise differences of natural economic regionalization are identical, no matter what their rank. Namely, objectivity, intrinsic interrelationship and socio-historic conditionality, the multi-faceted nature of regionalization, the arbitrary character of the boundaries, the "contour" character, monologism, and the appropriateness of utilizing comparable and identical indices in assessing the natural economic potential for the study districts.

Based on the aforementioned fundamental propositions, we have undertaken an integral natural economic regionalization effort for the Asian part of Russia in order to identify macro-regional natural economic systems.

As the main territorial consolidating factors for identifying the regions, we used the peculiar geological structural features of the study territory, its intrinsic transport-geographical location, the well-established socio-economic profile, and the natural and economic conditions of its transformation in prospect. Typically such regions have clearly pronounced axes and cores of district formation, and their system of pre-existing and eventual industries produces or must produce the key branches of 'captive' energy-production cycles of the heterogenous type. Large-size consortiums can be (or are already being) set up within their confines for the tapping or deep processing of the particular natural resources, independent of the boundaries of the country's administrative-economic formations.

Overall, for the Asian Russia we have identified none natural economic regions (see the figure), with a brief rationale of such identification explained later in the text. Numerical data on their natural and economic potentials, expressed in terms of arbitrary units calculated in accordance with summary physical indicators that were established for the entire territory of the Asian Russia, are presented in the table. The technique for calculating the arbitrary units for the natural-resource and economic potentials, respectively, can be found in a paper of I.L. Savelieva [4], and B.M. Ishmuratov [7].

Figure: Natural and economic regions of Asian Russia.

1. I-IX – region numbers: I – Ob-Irtysh, II – Kuznetsk-Altai, III – Northern Yenisei-Lena, IV – Angara-Yenisei, B – Baikalian, VI – Central-Amur, VII – Yana-Kolyma, VIII – Far Eastern Southern-Primorski, IX – Far Eastern Northern-Primorski; 2-4 Boundaries: 2 – Russian Federation, 3 – subjects of the Russian Federation, 4 – natural and economic regions of Asian Russia.

Table

Natural and economic potential of the natural and economic regions in Asian Russia as of the end of 2000

Number on map	Regions	Natural potential						Economic potential				Potential density, st.units/mln km^2		Gross social product, st.units
		Total, st.units	structure in species, %					total, st.units	structure, %			natural	economic	
			MR	FE	L	Ll	W		M	A	L			
I	Ob-Irtysh	116.0	1.8	44.8	23.4	15.6	14.4	85.6	30.0	49.2	20.8	59.0	4.4	27.4
II	Kuznetsk-Altai	82.7	17.0	37.3	31.4	4.8	8.9	64.9	55.8	35.1	9.1	15.4	12.1	22.9
III	North-Yenisei-Lena	83.0	30.8	1.0	0.5	14.6	47.1	5.0	4.2	2.8	93.0	14.2	0.2	3.6
IV	Angara-Yenisei	110.8	17.2	10.0	15.4	26.8	30.6	53.7	40.2	35.0	24.8	7.1	3.5	17.0
VI	Baikal	48.5	36.3	1.2	20.2	16.2	26.1	19.2	37.8	23.8	38.4	5.2	2.0	7.5
V	Central-Amur	51.3	7.1	5.6	12.4	19.8	55.1	15.2	26.0	26.4	47.6	3.9	1.2	3.9
VII	Yana-Kolyma	26.9	23.1	0.3	10.1	8.7	57.8	7.6	10.8	13.0	76.2	1.5	0.4	3.3
VIII	Far-Eastern Southern-Primorski	55.7	15.5	3.4	14.4	20.8	45.9	40.6	41.6	35.4	23.0	5.3	3.8	12.9
IX	Far-Eastern Northern-Primorski	25.1	12.4	0.2	9.3	16.1	62.0	8.2	12.8	29.2	58.0	2.1	0.7	1.5

Note. Kind of natural potential: MR mineral-raw materials, FE fuel-and energy, L- land, Ll logging-lumbering, W water, including hydropower; kind of economic potential: M manpower resources, A production fixed assets, L land.

1. **The Ob-Irtysh** region combines the Hanty-Mansi and Yamalo-Nenetsky Autonomous Districts, and the Tyumen, Omsk and Tomsk regions. Its main territorial-consolidating factors include the occurrence in conjunction with a series of highly productive oil and gas bearing structures of the West-Siberian Plate (West-Siberian oil and gas bearing province), the presence of clearly defined economic support axes of district formation: the navigable Ob river and the lower Irtysh, the Siberian railway main line with a number of sublatitudinal siding tracks, and a radial network of oil and gas pipelines. The city of Tomsk is a district-forming center. A basis for identifying the aforementioned subjects of the Russian Federation is also provided by the fact that they include a number of enterprises specializing in a deep processing of petroleum and gas resources (the Omsk and Tomsk integrated petrochemical plants, an extensive network of industries for processing casing-head gases: the Gubkinsky, Muravlenkovsky, Noyabrsky, Kogalymsky, Variegansky, Belozerny, Krasnoleninsky, Lokosovsky, Yuzhno-Balyksky, Surgutsky, and Megionsky plants; the Tobolsk integrated petrochemical plant (using, as raw materials, raw gas petrol supplied by casing-head gas processing plants), and the integrated oil and gas plants in the towns of Syagan and Surgut. This region boasts also sectors of the national economy that are beneficial for social and production infrastructural development of the oil and gas extraction fields that supply agricultural produce and forest resources that are tapped in an integrated highly efficient manner (the Tobolsk, Tomsk and Tyumen integrated logging-lumbering enterprises).
   The region is characterized by the highest (in the Asian part of Russia) capital-labor ratio, a low density of labor resources, with moderate labor productivity, and by the highest potential of fuel and energy resources (concentrating over 70% of the Russia-wide balance oil reserves, and about 80% of natural gas). The region has a country's lead in the tapping of these resources. It is also characterized by high potential of the forest resources. The land-resource potential is realized mostly through reindeer breeding and fur game hunting in the northern and middle parts of the region, and through agriculture and stockbreeding plus sheep husbandry and the use of game animals in the southern part. The structure of gross social product is dominated by industries (59.1%), while agricultural production accounts for 10.6%. As regards the structure of the main elements of the region's productive forces, it is placed into the category of raw-materials specific and capital-intensive ones, with the predominantly extensive utilization of the land resources.
2. **The Kuznetsk-Altai** region (the Novosibirsk and Kemerovo regions, Altai Territory, and the Republic of Altai) corresponds to the Sayan-

Altai mountain-folding system, with a large intermontane trough, and the southeastern part of the most elevated (with little prospects for oil and gas) West-Siberian Plate. The consolidating axis of district formation is provided by the Novosibirsk – Yurga – Kemerovo – Novokuznetsk – Barnaul – Novosibirsk railway ring, with a number of siding tracks. At the junctions of the ring with the main water arteries there emerged major centers specializing in a deep processing of products from the region's coal mines, ore mining and treating enterprises, timber industries, and agricultural undertakings, based on using local natural resources. Thus, through the use of coking coals from the Kuznetsk coal basin and their own iron ore resources of the Mountain Shoriya and Kuznetsk Alatau, as well as from East Siberia and from other regions of the country, it was possible to create a widely branching coal-energy-metallurgical-chemical cycle. It is formed, in addition to mining enterprises and concentrating mills, by the Novokuznetsk and West-Siberian integrated iron-and-steel works, the Guriev iron-and-steel works, the Novokuznetsk ferroalloy plant, the Kemerovo by-product coke plant, and by a number of thermal power stations. Altai Territory boasts a major complex of chemical industries, with the city of Barnaul constituting the center of territorial structure.

The region is characterized by the highest density of total natural-resource potential in the circumstances of the Asian Russia (154.5 standard units per 100 thous sq. km). Its structure is dominated by the fuel and energy component (37.3%). An important role in the development of the region's productive forces is also played by highly productive land resources, through an intense utilization of which it was possible to develop, on a large scale, grain-growing agriculture, beef and dairy animal husbandry, and fine-fleeced sheep farming, a basis for wide-scale development of food-processing industries to produce butter and cheese varieties. The mean potential density of intensively used lands for the region exceeds by a factor of three and five or six the figure for all Russia and its western part, respectively. According to the structure of the region's productive forces, the region refers to industrial-agrarian, labor-intensive regions, with a low capital-labor ratio and low labor productivity, and a high potential of intensively utilized lands and fuel and energy resources. The structure of gross output is dominated by the industrial component (56.6%), while agricultural production accounts for 15.7%.

3. **The Northern Yenisei-Lena** region includes the Extreme North of Krasnoyarsk Territory (the Dolgano-Nenets Autonomous District, and most of the Evenki Autonomous District), and western part of the Sakha Republic along the Lena river, and the northeastern part of

the Irkutsk region. The main reasons why the above-indicated territories were to be united into a common natural and economic region are: their location within the Siberian platform characterized by a specific complex of natural resources (oil and natural gas of the Yenisei-Anabar, Lena-Tunguska and Lena-Vilyui provinces; coal from the Tunguska and Lena fields, diamonds, iron ore deposits, and Iceland spar); the transport-geographical location; the well-established socio-economic profile; and the natural and economic conditions for its prospective transformation. The region's external transportation and economic relations are only possible via water and air, while the internal relations are maintained through all-terrain vehicles, as well as by air and fluvial transport.

This territory is characterized mainly by an extensive economy; more specifically, traditional reindeer-breeding and hunting, as well as fishery. The unique mineral deposits form the basis for industrial centers (both existing and to be established) built around intense utilization of natural resources, such as the Norilsk industrial center with a highly developed mining and metallurgical complex specializing in extraction, and pre- and post-processing of copper-nickel-rare metal ores; and the Mirny, Aikhal and Anabar diamond mining centers. The recent years saw the emergence of a number of oil and gas extraction complexes on the proven fields in the Baikit (Krasnoyarsk Terrutory), Vilyui (Irkutsk region, and the Republic of Sakha), Angara-Leena (Irkutsk region), and Vilyui (Republic of Sakha) oil and gas bearing territories of the Lena-Tunguska and Lena-Vilyui provinces; a solid basis has been created for the development of the country's large-scale oil production sector of the 21th century. It will be possible to satisfy not only the internal requirements of East Siberia and the Far east but also to enhance the export relations with the Asian-Pacific region's countries.. It is planned to create a unified system of oil and gas trunk pipelines.

Exceptionally high is the territory's water potential; however, its water-transport and fishery components have been exploited to date. The hydro power component is only also only partially utilized (the Vilyui, Khatanga and Kureika rivers). The same applies for hydroeconomics – in areas of emergent large-scale mining and metallurgical industries.

The region is distinguished by the lowest potential indicators of intensively used lands (0.07 standard units per 100 thous. sq. km), resource density (0.08 standard units), and fixed assets (0.07 standard units per 100 thous. sq. km), but by the highest labor productivity (by a factor of 4.6 higher than the Russia-average level). It is a region of highly productive industrial centers that have been and will be created on the basis of exploiting highly attractive resources and extensive background industries. The key role in the land structure is played by reindeer pasture lands and forest resources.

4. **The Angara-Yenisei** region includes the southern part of Krasnoyarsk Territory, the Republuics of Tyva and Khakasia, and the southern part of the Irkutsk region save the extreme eastern and northeastern parts, i.e. encompassing the areas of geological formations of Baikalian and Caledonian folding. The main territorial-consolidating factors are the water resources of the Angara and of the upper and middle Yenisei, convenient access as regards transportation and geographical location, relatively favorable natural conditions for human life and for the development of multi-sectoral agricultural production, and the wealth and diversity of forest, fuel and energy and mineral-raw material resources. A comparively extensively developed land transport system, combined with high water-transport, hydropower and water-economic potential of the rivers and with the availability of natural (land and forest) resources, has had a substantial influence on the structure of the industrial districts, clusters and centers specializing in the electric energy generation, as well as energy-intensive and water-intensive industries.

   The region is characterized by a shortage of manpower, has a moderate capital-labor ratio and moderate labor productivity, and is notable for high density of natural-resource potential (71.3 standard units per 100 thous. sq. km) against the average figure for the country's Asian part (46.7). As regards the potential density of forest resources, it heads the list in this part of the Russian state.

5. **The Baikal** region constitutes the Baikal natural area, comprising two ecological zones: the central zone that was declared a World Natural Heritage site, and the buffer zone. It includes the extreme eastern Baikal-watershed portion and the northeastern mountain-taiga part of the Irkutsk region, the Republic of Buryatia, and the western part, the Chita region. Thus the key territorial-consolidating factor is represented by Lake Baikal's watershed basin, and this imposes some ecological requirements on the economic activities here. On the other hand, it also is a single unit in the geological-structural context. It is the territory of development of Baikalian and early Caledonian folding, which predetermines the unity of mineral and raw material specialization; more specifically, the presence of a broad spectrum of precious, rare and noble metal deposits. The region's consolidation is also determined by a interrelationship of its transport-and-geographical location, a relative identity of its natural conditions for human life, an identity of the economic profile and its territorial structures, human settlement patterns of the same type,

and the mutually complementary and interrelated nature of the individual elements of the national economic complex. As was mentioned earlier, of special consolidating importance, however, is the location of these territories within the Lake Baikal watershed basin.

The region's economy has an agro-industrial profile – 46.6% and 16.2%, respectively, correspond to industries and agriculture in the structure of gross social product. The region is characterized by a shortage of manpower, a low capital-labor ratio, and increased labor productivity. Mineral resources head the list in the structure of total natural-resource potential. As regards the density of total potential (51.5 standard units per 100 thous. sq. km), the region is intermediate among the other regions of Asian Russia.

6. **The Central-Amur** region includes the southern part of the Sakha Republic (Yakutia), the Amur region, and the eastern part of the Chita region. The reason behind their unification into a common natural and economic entity was primarily their linkages as regards transport and economy. Transport operations are being carried out by means of the Baikal-Amur railroad and the Trans-Siberian railroad, and of the railway leg Bamovsky – Tynda – Berkakit for exporting products from the Aldan industrial center to Asian-Pacific countries. It includes two major geological-structural elements: the Aldan shield, the territory of which, upon completion of the railroad leg Bamovsky – Berkakit – Yakutsk, should likely be assigned to the Siberian platform, i.e. to the Northern Yenisei-Lena region, and to the region of development of Gerts and Caledonian folding. Today, within the Aldan shield, the mineral resources formed a basis for a major industrial center specializing in the extraction of gold and coking coals. According to the well-established international relations, coal produced there is to be exported mainly to Japan. Of important consolidating significance is also the linkages between existing and future economic systems in the southern and northern parts of the region: in the south, agrarian and industrial systems for processing agricultural and timber crudes, as well as machine building systems; in the north, industrial centers for lumbering, and gold, coal and rare metal mining. The region is also characterized by a high water-transport, hydropower and water-management potential of the Amur, Zeya and Bureya rivers. It is notable for a shortage of manpower, and an elevated capital-to-labor ratio and labor productivity.
7. **The Yana-Kolyma** region includes the eastern part of the Sakha Republic (Yakutia), and the Magadan region. Geologically and structurally, it is the Verkhnoyansk mountain system with a wealth of existing and potential mineral resources, with the primary

importance attached to gold, silver, tungsten, polymetals, rare and scattered elements, and quartz raw materials. The only way of communication between the region and Russia's outer districts and with foreign countries are water and air routes; internal integration is also possible via motor roads. It is a central industrial low-productivity region specializing in raw materials and a low density of manpower (ranking next to the Northern Yenisei-Lena region) which in the not too remote past had a high capital-to-labor ratio and labor productivity. The region is also distinguished by a very low potential of intensive utilized lands (0.56 standard units per 100 thous. sq. km), and by a high water potential, comprising mainly the water-transport, water-management and fishery components.

8. **The Far Eastern Southern-Primorski Krai** region includes Primorski Krai, Khabarovsk Krai, the Sakhalin region, and the Jewish Autonomous Region. The unification of these territories into a common natural and economic region was based on the promising economic and geographical location at the junction of the main land and sea transport routes, favorable for its external economic relations, as well as on its belonging to a common geological-structural element, the Sikhote-Alin folding system with the Urgalsky fporedeep. Its ore mining focus includes gold, tungsten, mercury, tin, and polymetals, together with its non-ore specialization: petroleum and natural gas of Sakhalin Island and the Sakhalin shelf, and coal from intermontane depressions. Based on tapping Sakhalin petroleum, it was possible to create the Khabarovsk and Komsomolsk-on-Amur refineries; international relations with Asian-Pacific countries are being intensively developed. Along with the identity of the natural-resource and transport-geographical potentials, the region is characterized by the same type of human settlement, and by the mutually complementary and interrelated character of the components of the economic complex.

   The region is notable for a shortage of manpower; and the figures of its capital-to-labor ratio and labor productivity are close to the Russia-average level. Its economy has an industrial character with a focus on raw materials.

9. **The Far Eastern Northern-Primorski** region includes the Kamchatka region, and the Koryak and Chukchi Autonomous Districts. Geological-structurally, it is the northwestern part of the Russian fold structures of the Pacific Cainozoic zone encompassing the Koryak upland, Kamchatka and the Kuril island arc, and Sakhalin island. Sakhalin belongs to the Far Eastern Southern-Primorski region according to its transport-economic relations. The region's main mineral resources are gold, sulfur, mercury, and precious and

rare metals. Its territorially consolidating factors are also the attraction of its transport-geographical location, the similar features of the established and eventual economic profile, and the unity of the ethnographic situation. The region's external economic relations are maintained by sea and air transport, and domestic links are sustained by all-terrain vehicles, and air and water transport. Of special significance for the region, as is the case with the Far eastern Southern-Primorski region, are the inland waters and the seas as regards the reserves of bioresources, with the leading position held by salmonid fishes and maritime animals. There exists a respective large-scale industrial processing sector.

As regards the structure of the elements of productive forces, it is a capital-intensive region with low labor productivity and an immense shortage of labor resources.

In closing, it would be appropriate to point out that natural and economic regionalization is essentially a problem-oriented issue. Its need is dictated by the ever increasing attention to the regional economic policy. Based on the specific characteristics of the units identified as regards natural resources and the economy, it is possible to formulate specific problem-oriented objectives to optimize the environmental conditions as well as structural organization of the production. The territorial natural and economic units thus identified open up a possibility of a comprehensive multisectoral development of the administratively heterogeneous territories without affecting in any way their political and administrative and economic sovereignty. The economic relations focused on a particular target within the regions thus identified must be sustained and regulated by associations, corporations, consortia, etc. which have been and gaining increasing momentum over the recent years.

REFERENCES

1. Saushkin Yu.G. Natural and economic regionalization of the Soviet Union. Ventn. MGU. Ser. 5. Geography 1980, No. 4.
2. Baransky N.N. The Making of the Soviet Economic Geography: Selected Works. Moscow: Mysl, 1980.
3. Savelieva I.L. Natural and resource regionalization, and its mapping. Geografiya i prirodnye resursy, 1993, No. 4.
4. Savelieva I.L. Natural and resource regionalization of Russia. Ibid., 1997, No. 1.
5. Komar I.V. Rational Utilization of Natural Resources, and Resource Cycles. Moscow: Nauka, 1975.
6. Kolosovsky N.N. The Theory of Economic Regionalization. Moscow: Mysl, 1969.
7. Ishmuratov B.M. The economic assessment map for agricultural land productivity in the south of Krasnoyarsk Krai (principles and technique of compilation). Dokl. In-ta geog. Sibiri i Dalnego Vostoka. Irkutsk, 1971, No. 32.

CHAPTER

11

Typological Analysis of the Degree of Siberia Development

L.A. BEZRUKOV

A scientifically grounded decision on the strategic guidelines for Siberia development calls for an objective assessment of the increasingly greater differentiation in the socio-economic state of separate parts of Russia's largest macro-region. When considering the transformation processes that are characteristic for Siberian regions at the post-Soviet stage, a special emphasis should be placed on the study of the influence of such basic economic-geographical factors as the degree of development and population. By the degree of development is meant the result of the process of development o a territory, i.e. characterized by the presence of the attained demographical and economic potential as the result of the inclusion of the territory in social turnover.

The use of subjects of the Russian Federation as estimation units should be considered inadequately justified, because in this case the extremely contrasting picture of physical distribution of the population and economy within large, heterogeneous Siberia regions is smoothed. The upper "storey" of a local (lower) level of administrative-territorial division (ATD) is best suited for identifying territorial characteristics of the degree of Siberia development, i.e. the level of administrative districts and cities under the republican, krai, regional and okrug jurisdiction which, by the way, is indeed recommended for such purposes by many outstanding economists-geographers [1, and other publications]. Moreover, it is mainly on the basis of such cities and districts that upper-level municipal formations are being created at present.

Administrative districts were treated in close association with cities and towns of a relevant status located within their confines, which implies the inclusion of them in the composition of districts. Besides, five territories were assigned to such intial cores, which are under the jurisdiction of the city administrations of Dudinka, Igarka, Neryungri, Norilsk, and Yakutsk. As a result, for the whole of Siberia (with the Tyumen region, and the Republic of Sakha (Yakutia)), a set of 400 cores (micro-districts and areas) was obtained, which are located within administrative districts (except for the aforementioned five cases) [2].

It must be specially stressed that the issue of unification of the districts and towns into common areas has a fundamental character. We operate on the premise that a city (especially a large and poly-functional one) will, to a significant extent, organize by itself the socio-economic space and has influence on the more remote territories; the city and the urban area surrounding it are in an inseparable unity and must therefore be regarded at a level of a micro-region jointly as a single whole. Predominantly on the basis of the same "nodal" principle of attraction of a rural area toward center-cities, Leizervich [3] identifies "economic micro-districts" of Russia. The logicality of the author's approach in this respect is also confirmed by the dynamical features of the number of lower administrative-territorial units of Siberia for the last 40 years. While in 1959 on the territory of Sibria there were 400 administrative districts and 58 cities and towns under the republican, krai, regional and okrug jurisdiction [4], in 1999 the number of districts even somewhat decreased, to 395, and the number of cities and town of the aforementioned status, on the contrary, increased to 124, i.e. by a factor of 2.1 [5]. Consequently, the change of the ATD was taking place solely through a growth of the towns and cities and an increase of their status, whereas the network of administrative districts was generally distinguished for a surprising stability.

For identifying the types of micro-districts from the degree of development, the underlying attribute (indicator) was represented by the population density, reflecting both the demographical potential and, to a significant extent, the level of economic development, as well as, in part, the type of economy, and the degree of environmental comfort for human living and activity [2]. As auxiliary typological attributes, our characterization use the indicators of the share of the urban population, the number and size of towns and cities, and the location of the areas relative to round-the-clock operating main traffic arteries, primarily railroads.

As a result of assessments based on statistical information about the population for the year 1999 [5], for the territory of Siberia it was possible to identify six types of micro-regions according to the degree of development (Fig.) which are combined into two main zones:

Figure: The types of Siberia's micro-regions according to the degree of development.

Degree of development (population density, pers per sq. km): 1 – high (more than 50), 2 – increased (25-30), 3 – medium (10-25), 4 – decreased (1-10), 5 – low (0.1-1), 6 – extremely low (less than 0.1).

a) the zone of continuous and arbitrarily continuous development (areas of the first and fourth types of the degree of development) located mainly in the southern and, partly, middle parts of a micro-region that is encompassed by a network of railroad trunk lines and year-round operating motor roads (totaling less than 1/3 of the surface area); and

b) the zone of sparse and core development (areas of the fifth and sixth types of the degree of development), occupying the sparsely populated taiga and tundra spaces of the Far and Near North, as well as, partly, localities of the southern mountain systems difficult of access (over 2/3 of the total surface area).

The first type of the degree of development is characteristic for a small number of highly urbanized micro-regions with a high degree of development and population density which develop on the basis of the cores of urban agglomerations, centers of a regional and sub-regional level (table). About 44% of all Siberia's population live in an area slightly exceeding 1% of its territory.

Table

General characteristic of the types of Siberia's micro-regions according to the degree of development

Type of micro-region (area)	Degree of economic development	Population density	Number of micro-regions (areas)		Territory		Population size	
		pers./km^2	units	%	thou km^2	%	thou pers.	%
First	High	More than 50	22	5.5	104.3	1.1	10989.2	43.8
Second	Increased	25-50	11	2.8	56.5	0.6	1588.9	6.3
Third	Medium	10-25	54	13.5	213.2	2.2	3090.6	12.3
Fourth	Decreased	1-10	236	59.0	2540.6	26.3	8111.5	32.3
Fifth	Low	0.1-1	55	13.7	3243.5	33.6	1192.7	4.7
Sixth	Extremely low	Less than 0.1	22	5.5	3494.2	36.2	151.0	0.6
Total for Siberia		2.6	400	100	9653.2	100	25123.9	100

The chief distinguishing feature of the type 1 areas is necessarily the presence in each of them of a significant center-city that wholly dominates over the adjacent territory and forms its main economic-settlement structures. A basis is typically provided by the largest (with a population over 1 million), major (250 thousand – 1 million) and large (100-250 thousand) cities; as an exception, medium-sized cities (classification of the cities according to [6]). Not counting the centers of autonomous okrugs (districts), of the other 14 regional centers, 11 districts are within the areas of the type under consideration (Novosibirslk, Omsk, Krasnoyarsk, Irkutsk, Barnaul, Tyumen, Kemerovo, Tomsk, Ulan-Ude, Yakutsk, and Gorno-Altaisk). A total of 36 cities and towns are situated here, including the two Siberian "millionaires" (Novosibirsk and Omsk), as well as 9 major cities (out of 12), and 9 large cities (out of 15). The proportion of the urban population is maximal and constitutes in these areas (with rare exception) 85-95% or higher, which is suggestive of their high urbanized status.

Micro-regions of the first type within the framework of a the central-peripheral dichotomy refer undeniably to central areas, performing the fullest set of organizational functions and characterized by a well-developed tertiary sector of the economy. The functions performed by their base cities are quite varied: almost all of them have the role of major economic centers (industrial, transport-distribution, commercial-warehousing, etc.); of important significance are the servicing, administrative-managerial, cultural, financial, market-infrastructural, and (more rarely) innovative scientific and informational functions. The overwhelming majority of the areas have a clearly pronounced industrial specialization, normally, with a very large-scale industrial production. In the surroundings of the urban centers and agglomerations, predominantly agriculture of the suburban type is pursued.

Almost all micro-regions under consideration are situated within the Main Zone of human settlement of Russia and are supported by railroad lines and year-round operating motor roads. The sole exception is, primarily, the Norilsk area in the north, having, nevertheless (thanks to the Norilsk-Dudinka railroad and the Dudinka-Murmansk route in the Arctic Ocean) virtually a year-round communication with the "Greater Land"; econdly, the Yakutsk area in the north, and the Maiminsky area (Gorno-Altaisk) in the south, having only motor road access to Siberia's railroad network.

The second type of the degree of development includes a very small number of urbanized micro-regions with an increased degree of development and an increased population density, which develop on the basis of a sub-regional and, partly, regional level, and in the peripheral zones of the agglomerations. Occupying 0.6% of Siberia's territory, areas of this type concentrate over 6% of its population.

As in the case of the areas of the preceding type, their characteristic feature necessarily is the presence of one dominant center-city (or, sometimes, several such centers at a time). However, base cities are now represented by less significant cities, mainly medium-sized and, more rarely, large cities. Of the cities situated here, Chita refers to major cities; there are 3 large cities, and 7 medium-sized cities. Their high urbanized status is attested by the impressive proportion of urban population that varies from area to area over the range of 75 to 90%.

Varied central functions are performed by regional centers only: Chita, Abakan, and Kyzyl, while the other areas with their cities and towns within the framework of the central-peripheral dichotomy can be referred, to a greater extent, to semi-peripheral districts. The overwhelming majority of the areas are distinguished by a clearly pronounced urbanization, with a high and moderate level of development of industrial production. Agriculture is mostly concentrated in suburban areas.

All micro-regions of the second type are situated inside of the Main Zone of human settlement of Russia and are supported both by railroads and by year-round motor roads. The only exception is the Kyzyl area that is connected with the railroad network of Siberia through a permanent motor road.

The third type of the degree of development corresponds to a rather large number of micro-regions of a medium degree of development and population density that have shaped themselves around local (more rarely, sub-regional) level centers along main railroad lines. Over 12% of Siberia's population live in an area occupying only 2.2% of its territory.

All micro-regions are situated within the Main Zone of human settlement of Russia and are supported by a network of railroads and motor roads. The role of base centers in more than a half of the areas is played largely by small cities (23) and, more rarely, by medium-sized cities; the sole exception is one major city (Bratsk), and one large city (Mezhdurechensk). Centers of the other areas are represented either by settlements of the urban type or by the most significant rural settlements (district centers). The degree of urbanization differs also: while in areas with developed industries and cities the proportion of the urban population reaches 80-90% (the Bratsk, Mezhdurechensk, Guriev, Sharypovo, Zima and other areas), in 12 West-Siberian micro-regions there is no urban population at all.

As regards the set of the functions performed (a limited character of the organizational potential) and specialization of the economy, areas of the third type are transitional from semi-peripheral to purely peripheral ones. The economic structure in about one half of the areas is dominated by industries whose level of development is high in a small number of them (the Bratsk, Sharypovo and other areas). In the other half of the areas, agricultures is of an increased significance.

The fourth type of the degree of development includes the greatest number of Siberian micro-regions which are characterized by a decreased degree of development and a decreased population density, as well as by their location around local-level centers both in near-railroad zones and in vast territories adjacent to them; they are supported by year-round motor roads. Areas of this type in the aggregate occupy more than one-fourth of the entire territory of Siberia and concentrate about one-third of its population, forming a continuous populated area stretching from west to east along the Trans-Siberian Railroad and its branch lines and encompassing, on all sides, micro-regions of the three preceding types.

Both in the south and mostly in the north, areas of the fourth type are beyond the conventional boundaries of the Main Zone of human settlement of Russia, which is indicative of the impressive progress in Siberia's development made over the last several decades of the 20th century (an expansion of the zone of continuous and arbitrarily continuous development). These areas have advanced especially further out to the north, occupying the petroleum and gas fields under intensive development in the Khanty-Mansi and Yamal-Menets okrugs (districts) of the Tyumen region. In addition to them, an isolated core of this degree of development has established on the basis of several areas in Central Yakutia which are distinguished by a rather dense rural population, with an attraction towards Yakutsk.

The functions of base centers in the areas are performed mainly by small cities (totaling 57), settlements of the urban type, and pivotal rural settlements (district centers). A similar role is played by the more significant cities here: one major city (Surgut), two large cities (Nizhnevartovsk and Ust-Ilimsk), and ten medium-sized cities (Nyagan, Nefteyugansk, Noyabrsk, Novy Urengoi, Kogalym, Kuibyshev, Lesosibirsk, Tulun, Ust-Kutm and Neryungri). The degree of urbanization is different: in areas with significant cities and large-scale industry, the proportion of the urban population reaches 80-90%, whereas in most areas there is no urban population as such. The smaller part of the areas has a clearly pronounced industrial profile; yet, a high level of industrial development is characteristic for a few of them (the Surgut, Ust-Ilim, Nizhnevartovsk, and other areas). Most of the areas are dominated by agriculture of different kinds. The taiga zone boasts its primordial primary pursuits, namely, hunting and fishery, as well as a variety of forest-related activities (lumbering, gathering of forest fruits, etc.).

The fifth type of the degree of development is characteristic for a rather large number of micro-regions with a low degree of development and a low population density. As these areas are remote from the railroad lines (with rare exception) and (mostly) from year-round motor roads, very vast territories still remain difficult of access via transportation, with only seasonal water ways and winter motor roads available, as well as too expensive air transport. The areas under consideration are all outside the limits of the Main Zone of human settlement of Russia and are surrounding the preceding areas (or the zone of continuous or arbitrarily continuous development) in the form of a continuous area on the northern side (the middle and northern taiga; in some places the tundra, and forest tundra), and occasionally on the southern side (the most remote, sparsely populated mountain-taiga and mountain-steppe areas stretching from the Altai to the Trans-Baikalia). Whilst micro-regions of this type occupy, in the aggregate, over 1/3 of Siberia's territory, less than 5% of its population live here. Almost all areas are officially assigned to the regions of the Far North and to similar localities.

The urban network is poorly developed: the 17 cities here are all small ones (Igarka, Dudinka, Kodinsk, Kirensk, Bodaibo, Severobaikalsk, Mirny, Aldan, Olekminsk, Verkhoyansk, and others). The functions of local centers are mostly performed by settlements of the urban type, and core rural settlements (district centers). The overwhelming majority of the areas is therefore distinguished for the absence or a small proportion of the urban population.

According to all indicators, areas of the fifth type represent clearly pronounced peripheral areas. The industries that are available in some of the areas are characterized mainly by a narrow mining and, partly, forestry specialization. In the taiga zone, along the valleys of large rivers, there are cores of animal husbandry and agriculture, in combination with native hunting and fishery and, partly, transport reindeer breeding (this is particularly true for aboriginal peoples); in the mountainous areas of the southern part – mountain-pasture animal husbandry pursued by aboriginal peoples, with some cores of agriculture, hunting, and gathering. Overall, the areas of this type have similar external attributes: harsh climate, sparse network of settlements, low population density, vast inhabited spaces, poorly developed road network, and largely extensive kinds of economy.

The sixth type of the degree of development includes a small number of micro-regions with an extremely low degree of development and population density, at large distances from railroad lines and year-round motor roads; they are connected with the "Greater Land" (not counting the expensive air transport) only at a limited period of navigation on rivers and the Arctic Ocean (Northern

Sea Route). The areas all refer to so-called regions of the Far North and encompass vast spaces of the tundra, forest-tundra and northern taiga, from Yamal in the west to Kolyma in the east (mostly the areas of the Yamal-Nenets, Taimyr and Evenki okrugs, and of Yakutia). It is the most extensive type as regards its area – over 36% of Siberia's territory; at the ame time it is the least populated, with only 0.6% of the population. Indeed, the population density here is nearly by a factor of 2700 lower than in the areas of the first type.

Areas of the sixth type exemplify the most backward and abandoned, deeply peripheral regions with extreme climatic conditions, an exceptionally sparse network of settlements, and immense uninhabited spaces, total absence of roads, and extensive kinds of economy. The urban network is virtually nonexistent – there is only a single small town (Srednekolymsk), and a few settlements of the urban type; therefore, the population lives mostly in rural areas. Industry is of a minor nature and has a strictly local significance (fish processing, and the like). Traditional primary kinds of economic activities: reindeer breeding, hunting, and fishery are very common pursuits, especially of aboriginal peoples.

The above typology of the micro-regions unravels rather adequately, we believe, the result of a joint effect of some factors that are responsible for the key territorial features (gradients) of development and settlement of Siberia. Traditionally, one such feature is the latitudinal zonality (the "south-north" gradient), based on the general tendency toward a decrease of the degree of development and settlement father out to the north (and, partly, to the south) against Main Zone of settlement which is the most favorable for living and economic activities. The other feature is the west-eastward asymmetry (the "west-east" gradient) implying in general a decrease of the rate of development and settlement in the eastward direction, which is historically associated with a gradual attenuation of the colonization "waves" originating from the country's European part and, partly, with an enhancement of the severity of climatic conditions in the eastward direction. The third feature is the clearly pronounced attraction of the economy and population toward the main traffic arteries (the gradient based on the "near-trunk route – areas difficult of access" dichotomy, which reflects the tendency toward an abrupt decrease of the degree of development and settlement with the distance from the trunk routes. And, finally, the fourth feature, the high concentration of population and economy around the key towns and urban agglomerations (the gradient along the "center-periphery" interaction line), unveils the patterns of attenuation inherent in the magnitude of influence of the largest centers with the distance from them (a decrease of the degree of development with the distance in peripheral regions).

The first two territorial features are clearly manifested mainly at a macro-regional (Siberia-wide) and regional level. They have been much studied over the past several decades and, for instance, were visualized in the now "classic" practices of identifying the zones of population concentration and economic development of Siberia as done by Pomus [7, 8].

The last two territorial features are also quite obvious at an intra-territorial and local level. They have been much more poorly investigated; this is especially true of current Siberian data, although an interesting fundamental publication on the southern part of West Siberia by Fuks [9] deserves mention. Our identified level of micro-regions opens up some promising opportunities for detailed study of these features (gradients) for the whole of Siberia. It is the above-indicated gradients that now have, perhaps, the best explanatory capability of a territorial differentiation of the degree of development and settlement, and it is they that largely provide the basis for our developed typology.

It should be pointed out in conclusion that under the Siberian circumstances, the standard market mechanisms alone are unable to provide accumulation, redistribution and rational usage of the surplus product for purposes of an equalization of the action of unfavorable factors and extremely sharp socio-economic contrasts. Therefore the existing clearly pronounced territorial differentiation of the degree of development and the related acuteness of the socio-economic situation calls for an efficient regional policy at a governmental level. This policy must rely on detailed research on the lowest possible territorial level – this paper is a summary of preliminary results along this line of enquiry. The suggested typology of Siberian micro-regions according to the degree of development and population density can serve both as a classification of territories depending on the socio-economic effect they yield (a decrease in the effect from the first to sixth type) and as the framework for the purposeful step-like switch-over the areas of different types to the market mechanisms of economy regulation (from the first to sixth type). Implementation of this technology in the managerial practice is made easier if it is adjusted to the existing lower ATD and, accordingly, to the system of upper-level municipal formations.

References

1. Pokshishevsky V.V. Geography of the USSR Population. Economic-Geographical Essays. Moscow: Prosveshcheniye, 1971, 174 p.
2. Bezrukov L.A. The administrative-territorial division and the degree of economic development of Siberia at the turn of the 20th century. The Administrative and State-Legal Development of Siberia of the 17th-21st Centuries. Irkutsk: MION-IGU, 2003, pp. 202-232.
3. Leizerovich E.E. Economic Micro-regions of Russia (Grid and Typology). Moscow: Trilobit, 2004, 132 p.
4. The RSFSR Population. Results of the All-Union Population Census as of 15 January 1959. Moscow: CSA at RSFSR CM, 1960, 239 p.
5. The Population of the Russian Federation in Cities, Settlements of the Urban Type and Districts as of 1 January 1999. Moscow: Goskomstat Rossii, 1999, 203 p.
6. The Urban Development Code of the Russian Federation. Rossiiskaya gazeta, 1998, 14 May.
7. East Siberia. The Economic-Geographical Description. Moscow: Geografgiz, 1963, 896 p.
8. West Siberia. The Economic-Geographical Description. Moscow: Geografgiz, 1963, 488 p.
9. Fuks L.P. Population Distribution in West Siberia: Self-Organization and Management, Results, and Problems. Novosibirsk: Izd-vo PRO, 2003, 216 p.

CHAPTER

12

Problems of Population Geography and Human Ecology in Asian Russia

N.V. VOROBYEV, K.N. MISEVICH AND S.V. RYASHCHENKO

Long-term research conducted by the B.V. Sochava Institute of Geography, SB RAS has aimed at laying down a geographical groundwork for the regional socio-ecological analysis. This is defined as a system of methods and approaches enabling an exploration of regional patterns of demographical, medical and social parameters characterizing population quality depending on ecological-geographical conditions of human environment. The "socium" (a territorial population entity) becomes the system's pivotal point and the subject of its assessment.

Socio-demographical and ecological investigations involve primarily looking for connections and dependencies between complex indices characterizing the population and its health, on the one hand, and the comfortableness of natural conditions in populated areas and pollution, on the other. The logical analysis chain appears as follows: population – settlement pattern – environmental comfort and pollution – integral estimation of the socio-ecological situation [1].

A major research step is the specification of the conceptual base of a socio-ecological analysis. It relies on such concepts as "territorial activities", "medical-geographical situation", "social health", "population quality", "socio-ecological system" etc. Methodological elaboration is required for the estimation criteria of demographical, medical, social and other components of the population's activities [2].

Analysis results are generalized in socio-ecological regionalization relying on assessments of the socio-ecological situation based on conjugated mapping, sector mapping and the typology of the processes under study. The final stage of the research provides complete demographical maps and a summarized demo-ecological-zones map for Asian Russia and its specific parts.

Based on the available Russian research experience in studying the role of (natural, social, economical, ecological) environments of human activities in shaping human requirements and the possibilities of their fulfillment, we estimated the above factors over each region in Asian Russia. The dependence of a person's life on the factors of their geographical and social environment are determined by the necessity to satisfy their everyday needs in food, health and safe (comfortable) environment, adequate defense level in conditions of discomfort, including the social, economical, recreational aspects of their life.

DEMOGRAPHICAL SITUATION AND THE REPRODUCTION OF THE POPULATION

The demographical development of Asian Russia in the 20th century involved primarily its settlement and economic development. This was accompanied by high population-growth rates, the emergence of a network of settlements, and by the rise of new cities and towns. Agricultural and industrial production grew, and new lines of communication were installed, along with a certain social infrastructure [3].

Over the past decade, however, the situation in the region has undergone dramatic changes. The natural population growth-rate has fallen several-fold, with an increasing natural decline, increased migration, a reduction in population -this is especially true of children under 16, and an increased proportion of old-age persons, whereas the life expectancy has decreased. To resolve the problems of regional demographic development, the recent dramatic situation in the east must be recognized taking into account the geopolitical and geo-economic situation, the territory's extremely uneven population, the sparse transportation network, especially in northern areas, the considerable natural geographical differences, the economic specialization in raw materials, and the population's socio-economical welfare and health.

The regionally uneven economic and social development urgently call for an improvement of the conditions of life. Unless the most general administrative, economic, social, and ecological problems are addressed, it is impossible, in the foreseeable future, to resolve the demographical problems. After all, it is the social orientation of the economy that is the most important goal of reorganizing the structural and investment policy in our country.

It is the regions of Asian Russia where the bulk of the nation's natural-resource potential is concentrated. The most diverse natural wealth can be found there: fuel and energy, ores and non-metallic mineral resources, water, forest, land, and other resources. All sorts of mineral resources utilized in the modern economy were discovered in this territory. That is why it is in the interests of the nation to rationally tap these resources and settle Russia's eastern territories.

On the other hand, the huge total area of Asian Russia, amounting to 12.8 mln km^2, or 74.7 % of the country's territory, its scattered population, and the severe natural and climatic conditions were the reasons for the differing levels of economic and social development in different regions, and for the differentiated character of territorial organization of production and population distribution. The population living there (30.8 mln, or 21.4 % of the country's total, as of 1/01/2002) have to bear on their shoulders a heavy burden of exploiting the natural resources and developing the economy.

The formation process of the Asian Russia's population has progressed in compliance with general laws of social development. It formed over a long time inseparably linked to the economic development of the territory. Its settlement pattern during particular historical stages fairly precisely reflected the specific aspects of the territory's economic development and the predominant exploitation of either a single natural resource or a complex combination of several resources, depending on the people's needs and requirements, business interests and national policy.

The most dramatic shifts in the economic development of the study area occurred during the post-war years, accompanied by an increase in population. Thus, from 1959 to 1992, Asian Russia's population grew by 9.9 mln, or 43.1 % (Table 1).

The population of the region grew through natural growth as well as 'mechanical' influx.

Until 1992, the industrial development concurrent with development of Siberia's and the Far-Eastern territory proceeded largely individually, in each sector of the economy. Large factories were built and provided with social and household infrastructural facilities. High coefficients were set on wages in order to attract manpower.

Table 1

Population dynamics in Asian Russia

Regions	Total population, thou			Increase in population, thou	
	1959	1992	2005	1959–1992	1992–2005
West Siberia	11252	15132	14678	3880	- 454
East Siberia	6473	9257	8424	2784	- 833
Far East	4834	8042	6593	3208	- 1449
Asian Russia	22559	32431	29695	9872	- 2736

Over the last several years (from 1992 on), as a result of changes in the domestic political and economic situation, the region's socio-economic development has suffered dramatic blows. Output has been declining, economic links were severed, transportation costs increased sharply, which had an adverse effect on the Siberian factories that are scattered over an enormous territory. The social sphere in many towns began to dwindle or it even disappeared. This resulted in a decline in living standards and, hence, in a decrease in birth rates, and an increase of the death rate, which was accompanied by increased migration outflows from a number of areas and towns.

Over the period 1992 – 2005, the region's population reduced by 2,736 thou (see Table 1).

An examination of the role played by various combinations of natural and migration-induced movement of the population, based on a well-known technique for analyzing the population dynamics [4], revealed a number of characteristic properties of population dynamics for the transition period 1992-2002. While the Soviet times involved only the types of dynamics with a natural population growth, all possible types can be observed to date [3]. Of the eight theoretically available types of dynamics at a level of regions, subjects of the Russian Federation, six types were identified, and type 3 and 4 characterized by a large migration-caused increase of the population were absent (Fig. 1). The prevailing types are 5 and 7, with a reduction of the population, but its components are different: the former case showed a natural decline not compensated by the migration influx, and the latter case exhibited a migration-induced outflow aggravated by a natural population loss.

The population was increasing only in four regions; accordingly, a reduction took place in 25 regions. While a natural growth was observed in 10 regions (national autonomies, including two oil and gas bearing districts), and a loss was in 19 regions, the migration influx of the population was observed in 7 regions (only West-Siberian), and the outflow was in 22 regions.

The population dynamics is influenced by a variety of factors, the most important of which are: the geographic location of the regions, characteristics of the socio-economic situation, and the ethnical and age structure of population. A number of subjects of the Russian Federation, sharing a common typological population dynamics, can be combined into territorial groups, i.e. districts of population dynamics.

Thus, the southeastern district includes Baikal's territories and all (except fro northeastern) Far-Eastern territories, characterized by the type 7 dynamics implying a migration outflow of the population exceeding many times a natural loss. In the 1990s, the Trans-Baikalia (Buryatia, and the Chita region) moved from type 8 to 7.

The southwestern district of Siberia, five regions of which are dominated by the type 5 dynamics, which is typical of all West Siberia as a whole. The Omsk region stands somewhat apart – it shows a migration loss of the population. The middle geographical location of Krasnoyarsk territory and the Republic of Khakasia is responsible for their intermediate position (type 6) between the eastern and western Siberian territories, having the type 7 and 5 type dynamics, respectively.

The northwestern district of Siberia inhabited by a relatively young (or recently formed) population in the Khanty-Mansi and Yamal-Nenets okrugs, is characterized by the type 2 and 1 dynamics, respectively.

The northeastern district, including the territories of national autonomies, with extreme natural conditions for life activities of the new arriving population, shows the type 8 dynamics.

Although the national regions of southern Siberia (Republic of Altai, and Republic of Tyva, and the Ust-Ordynsky and Aginsky Buryat Autonomous Okrugs) do not constitute a common continuous territory and have the type 1 and 8 dynamics, constitute a special group of territories with a natural population growth.

The analysis made in this study, and also the typology and identification of districts (territorial groups) give an insight into major regional differences of the population dynamics, and into the difficulties in using common approaches to resolving socio-demographic problems under Siberian circumstances.

Thus, over the past decade, Asian Russia with its growing population has transformed into an area of an ever dwindling demographical mass. The processes of settlement and economic development have changed sign. A long period of population being formed thanks to expanded natural growth and migration influx has given way to a reduced reproduction, and by the migration outflow.

REGIONAL MIGRATION FLOWS

The migration situation in Asian Russia's regions is evolving at the turn of the centuries, under a simultaneous influence of heterogeneous and oppositely directed factors of population movements. The regional development that is understood as many-sided, directional, regular changes in territorial systems, is the chief force 'organizing' the migration processes. The following major groups of migration factors are characteristic for the Siberian and Far-Eastern areas:

a) constant (uncontrolled) natural-geographical characteristics of the territory that have influence on migration as the degree of favorable living conditions when compared to the other territories;
b) historical heritage of the Soviet era of development of the country's eastern part with an artificial growth of the economy of many Siberian territories in the interests of the country's unified economic complex; the aftermath of this growth – such as the existing economic structure, the population pattern, and a set of life-support conditions – will persist over a long-term period;
c) the post-Soviet reform properties that have already created new qualities in the socio-economic sphere in the form of employment opportunities, career promotion acquisition of income, etc.

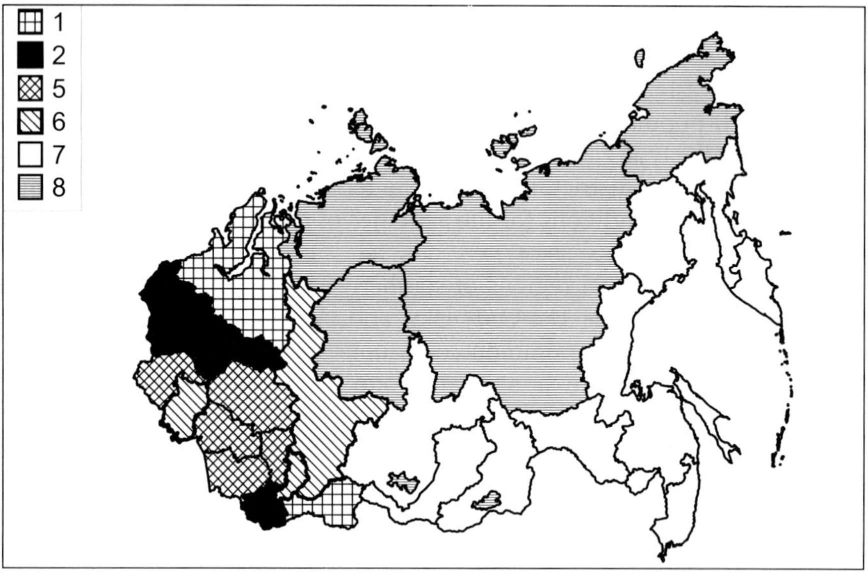

Figure 1: Types of population dynamics of Asian Russia's regions (1992-2002).

1 – growth: natural growth exceeds migration outflow; 2 – growth: natural growth exceeds migration influx; 5 – reduction: natural loss exceeds migration influx; 6 – reduction: natural loss exceeds migration outflow; 7 – reduction: migration outflow exceeds natural loss; 8 – reduction: migration outflow exceeds natural growth.

A true migration analysis combines three territorial levels:

- Asian Russia (for examining the general migration situation);
- the Baikal region (to provide better insight into the migration links of a inter- and intra-regional level); and
- the Irkutsk region (for detailing intra-regional migrations).

The migration total for Asian Russia has been negative in the past decade. In a regional setting, the migration patterns are not as unambiguous. The areas of the country's north-east underwent a particularly dramatic migration loss. The total losses of the country's east are manifest when compared to the migration situation in the west. All West-Siberian areas, with the exception of the Omsk region and Yamal-Nenets okrug, enjoyed population influxes. In East Siberia, it is only the Republic of Khakasia that had a population influx. As for the Russian Far East, all regions are migration donors. In view of the fact that Asian Russia's population inhabits primarily a narrow strip along the Trans-Siberian trunk railway, the type of Trans-Siberian migration profile is of some interest, allowing one to compare the migration outcome in the southern areas (Fig. 2). The diagram shows (left-to-right) a continuous west-eastward reduction in the relative migration growth (8 persons, on the average, from region to region).

Migration links with the CIS and Baltic states provide a considerable migration growth. West Siberia has particularly close contacts with neighboring Kazakhstan which 'donates' immigrants on a regular basis. East Siberia, with its migration influx from CIS countries, is very poorly tied with CIS countries, even in comparison with the Russian Far East, enjoying an almost balanced population exchange.

Of significant importance for Russia's Siberian and Far-Eastern regions are the migration ties with CIS and Baltic states, as they provide growth largely through Asian states; the European countries, however (such as the Ukraine) are recruiting the Siberian population.

The correlation of oppositely directed migration flows (series of arrivals and departures) is are similar in general. Thus, 1218 left Asian Russia per 1000 newcomers from the Central Russian regions. However, departures for foreign countries except CIS countries' exceed the arrivals from there bў more than 26 times, while the population departure for CIS countries is 2.5 times lower than the population influx from them.

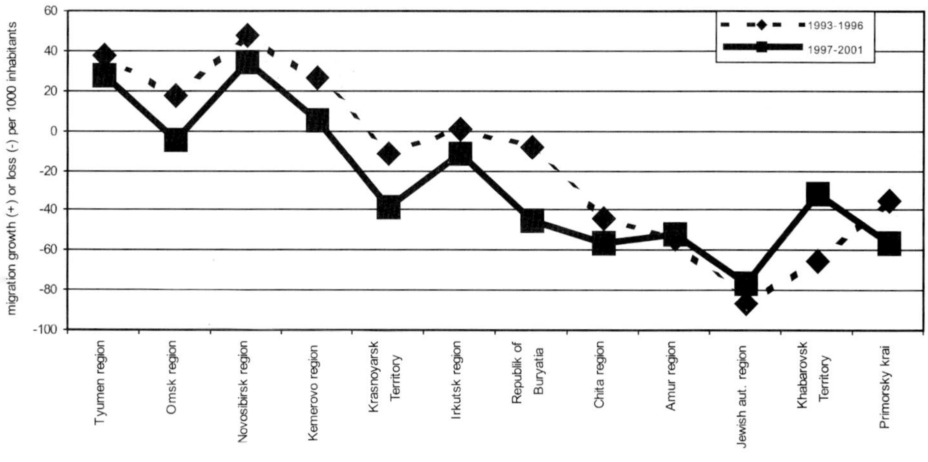

Figure 2: Mean annual migration changes in the population inhabiting the areas along the Trans-Siberian railway.

Migration ties with the Ukraine are more pronounced in the Tymen region, accounting for one-third of migration turnover. A significant influx of population corresponds to the Khanty-Mansi and Yamal-Nenets autonomous okrugs. As far as the other eastern regions are concerned, a sizable contribution to the Russian-Ukrainian migrations is made by Krasnoyarsk Trritory, Primorski Kray and

Khabarovsk Territory, the Irkutsk and Kemerovo regions, and the Republic of Sakha (Yakutia), yet they all undergo a population loss.

According to the census data for the year 2002, the proportion of the those born in the Ukraine is the highest in the northern and eastern parts of Asian Russia, with an almost immigrant population. In southern Siberia with its numerous native population, the share of those from the Ukraine is low, 2-3% of the population of the regions, which corresponds with the mean Russian level. The Russian-Ukrainian migration ties are influenced by a great variety of factors, such as the life-support system, the historically established population distribution, good employment opportunities, the level of wages, and the living standards.

That the European and Asian parts of Russia are relatively attractive is indicated by the oppositely directed migration flows. Thus, in the European Russia, of 1000 immigrants arriving from CIS countries, 321 persons departed to their respective countries; 684 departing persons correspond to Asian Russia; 611 and 1137 persons, respectively, correspond to the migration exchange with the Ukraine. Exchanges with the Ukraine have for Siberia and the Far east the same character as the intra-Russian migration exchange, where 1236 persons out of 1000 arriving depart again to their respective regions.

Inter-governmental migration of Siberia's and Far-eastern population is an indicator by which to infer the level, direction and rates of regional development. Thus, the migration aspect of depressiveness is expressed in terms of the population's relative migration passivity, retardment of its rural-urban redistribution, and most importantly, a constancy of the migration outflow beyond the territory.

Migration ties with foreign countries except for CIS states are highly selective as regards the migrants' country and their ethnicity. Most migration losses correspond primarily by regions with an abundant German population (the Omsk and Novosibirsk regions, and Altai Territory). Both absolute and relative emigration from East Siberia and the Far East is lower than that from West Siberia.

Russian internal redistribution of population results in migration losses which rises dramatically eastwards. Thus, over the period 2000–2001 only, the country's Asian part lost 101 thou, who departed for the European part, while West Siberia received 1 thou. Inhabitants, and with East Siberia and the Far East lost 34 and 68 thou., respectively. The list of donor regions is headed by The Republic of Sakha (Yakutia), Primorsky kray and the Chita region, each donating 6–7 inhabitants to Russia's other regions. Russia's internal population movement has a spontaneously natural character. The general trend today implies that the population moves to localities with more favorable living conditions, with a minimum possible cost of living. As the country stretches longitudinally, the geographic longitude has a more dramatic influence (compared with the latitude) the complex of living conditions and, hence, population migration.

Russia's internal population movements have mainly acquired an intra-regional character, as can be seen from migrations within and between specific federal districts. The Siberian Federal District is the most closed: 86.5 % of all immigrants and 81.9 % of all emigrants have moved within the region. In the process, migration movements prevail within regions, the subjects of the Russian Federation. Thus, intra-regional migration turnover constitutes 60.4 %, whereas the Republic of Buryatia, and the Irkutsk and Chita regions give 59.9, 61.9, and 62.9 %, respectively. They do not stand out among the Siberian regions like the 'super-closed' Tyva, where 82.9 % of all migration occurs within the Republic, or like the 'open' Republics of Altai, Khakasia, and the Tomsk region, where internal movements account for about 53 % of all migration turnover.

The relative intensity of intra- and inter-regional migrations is manifestly revealed by special factors which eliminate the influence of the size of population in the source and target territories. The relocation rate of population within the federal districts of eastern Russia (the Far Eastern, Siberia, and the Urals) exceeds that of inter-regional relocations by one order of magnitude or more.

The population redistribution between the components the Baikal region is fairly active, involving several thousand people annually. Thus, over the period 2000–2002, 8.4 thou arrived in the Irkutsk region from the Republic of Buryatia, and 5.4 thou from the Chita region. The same period saw 5.3 thou inhabitants leaving the Irkutsk region for the Republic of Buryatia, and 3.1 thou for the Chita region. The migration exchange with the Transbaikalia provided the Prebaikalia with a migration growth above 5 thou. The Transbaikalia accounts for 27 % of all arrivals from without the Irkutsk region, and for 14 % of all departures. With the share of the Transbaikalia in Russia's population constituting 1.5 %, the relative rate of these migration ties exceeds by one order of magnitude the average rate of the other migration ties. Statistical data show that the territories occupied by the Baikal region are closely tied by migration flows (Table 2).

Population migration losses in the Baikal region are aggravated by unfavorable structural patterns of migration flows. The population outflows are largely due to younger age groups; moreover, the level of education received by the migrants is higher when compared with the remaining population.

Let us examine intra-regional population migrations using, as an example, the Irkutsk region as a typical Siberian region with average migration characteristics. In this case we do not observed any extreme attributed inherent in the Far-Eastern territories, with the population departing at a large scale from them, on the one hand, and the southern West-Siberia fed by a human influx from Kazakhstan. The study revealed a migration pattern for the last decade separately for three territorial sections.

The 'town – village' migration section. Towns are experiencing a population influx, only one third of which is accounted for by urban areas, while two thirds are supplied by CIS countries; urban residents usually depart for the European Russia.

Table 2

Balance of migration flows in the Baikal region in 2001–2002*

		Territories of arrival					
		Irkutsk region	Republic of Buryatia	Chita region	Total for Baikal region	Other Russian regions	Total for Russia
Terri-tories of depar-ture	Irkutsk region	58377 39.6	3398 6.1	1940 2.9	63715 23.6	30529 0.4	94244 1.2
	Republic of Buryatia	5380 9.7	23434 112.7	3140 12.4	31954 31.5	12278 0.4	44232 1.5
	Chita region	3318 4.9	3064 12.1	33306 108.8	39688 32.3	19301 0.6	58989 1.7
	Total for Baikal region	67075 24.9	29896 29.5	38386 31.2	135357 27.4	62108 0.4	197465 1.4
	Other Russian regions	19031 0.3	4529 0.2	7800 0.2	31360 0.2	3927590 ...	3959819 1.0
	Total for Russia	86106 1.1	35194 1.2	46286 1.3	167586 1.2	3989698 1.0	4157284 1.0

1. * numerator: migration; denominator: relative intensity coefficient of migration links (1.0 is average for Russia).

The 'north – south' migration section. With a positive net balance of migration in the southern parts of the region, the migration outflow from the northern areas has exceeded it, notably, while the southward movements were not high (as low as a third of the outflow from the north), the external exchange of population has provided the south with more inhabitants than the northern lost.

The 'long-settled areas – recently settled areas' migration section. In a situation where almost all long-settled areas retain their population, a number of territories of recent pioneering settlement (the Kazachinsk-Lena, Nizhneilimsk, Ust-Ilimsk and Ust-Kut districts*) have lost a sizeable portion (up to 10 %) of their relatively young and recently-formed population.

Studies of the territorial differentiation of the migration situation in the Irkutsk region have shown that it was determined by a set of socio-geographical factors, which allowed four zone-typological groups to be identified within the local population:

a) the population of the southern, highly-urbanized, area along the trunk railway, characterized by low mobility with predominant intra-regional relocations, and by migration inflows;

b) the population of the region's middle areas, with moderate mobility, balanced counter-migrations, and highly closed (two thirds) migrations within the region;
c) the northern population, relatively recently formed, showing higher mobility, with the outflow by a factor of 1.5 higher than the influx; and
d) the population of the Far North, with a high rate of departing for the southern regions [5].

Assessment of the migration prospects yields unfavorable figures for Asian Russia in general. Migration shifts of the past decade are contrary to the trends observed for several centuries of Russian history. Thus, the population that formed in northern localities during the last quarter of the century in connection with geological exploration, construction , industry, and transport, lost their jobs in many locations due to the collapse of a considerable portion of factories and other facilities of these sectors of the economy. Opportunities to settle down for good in the northern conditions are negligible for the new-comers. Industries depend on external sources (the world and Russian markets, and support by the state). Agriculture has no commercial character and is risky. Hunting as a trade is limited and requires vast areas as well as special habits and skills in nature management. Therefore, only some of the northern population can live in the self-dependence mode: those employed in not numerous efficient enterprises taking advantage of current sales opportunities; those providing services to the infrastructure that is necessary for operating enterprises; employees of the budgetary sphere, and persons connected with local lifestyle of traditional nature management.

The problem of reversing the migration flows within Russia towards its eastern regions is complicated by several critical factors:

a) poor quality of life, and inadequate productive employment;
b) lack of funding in order to remedy these deficiencies;
c) lack of mobile population reserves in European Russia; and
d) lack of the political will to rectify the situation using public policy measures.

The prospects of migration lean upon the uncertainty of socio-economic development in the country's eastern macro-region whose attractiveness for the population can be achieved only through systemic changes in such spheres as the economy, life standards, high technology, and investments.

REGIONAL PROBLEMS OF HUMAN ECOLOGY

A key socio-geographical research area involves the study of the socio-demographic processes, and biosocial characteristics of population within the context of links and relations with the particular geographical environments of life activity (natural and technogenic).

According to existing medical-geographical research results on the territory of the Asian part of Russia, climate is the main factor of comfort at this level. We identified the following types of territory: absolutely extreme: absolutely extreme – Arctic coast and isles; extreme (subarctic subzone); extremely uncomfortable – northern-taiga subzone; and uncomfortable – middle-taiga subzone and semi-arid areas in Siberia's extreme south; moderately uncomfortable – south-taiga subzone and arid-steppe regions of the Transbaikalia; the most comfortable – flat forest-steppe areas of Western Siberia, forest-steppe near-mountain areas, areas of large troughs in East Siberia, and broad-leaved forests in the Far East [6].

Table 3

Population distribution in territories with differing comfort levels of natural conditions

Comfort of natural conditions	Percentage of stable population
Abs. extreme	0.01
Extreme	0.4
Sub-extreme	3.0
Highly uncomfortable	4.6
Uncomfortable: northern	12.8
Southern	3.8
Moderately comfortable:	
northern	11.9
southern	12.8
Most comfortable	50.6

The main concentration centers of the population are Siberia's industrial districts. Virtually about 70 % of Siberia's population live in the industrially modified environment, which, in the Siberian setting, is becoming of decisive importance for the quality of human life. About 50 % of West Siberia's population and nearly 70 % of East Siberia's population live in heavily contaminated environments (Table 4).

Table 4

Population distribution in territories with differing environmental pollution levels, %

Pollution level	West Siberia	East Siberia	Far East
Maximum	47.1	68.4	-
Medium	51.7	9.5	52.1
Low	0.7	19.9	38.1
Minimum	0.5	2.2	9.8

The overwhelming majority of administrative krays and regions are facing a population decline, with a negative natural growth. The only exceptions are the national entities: the Republics of Altai, Sakha, Tyva, and the Chukotka Autonomous District.

Processes that have received the name 'depopulation' of territories, as well as of Russia in general, are the result of the high death rates against low birth rates. The highest death rates are not characteristic of areas with uncomfortable natural conditions, rather they are typical of areas with high population densities and industrial concentration (the Kemerovo, Novosibirsk and Irkutsk regions, and Krasnoyarsk territory), which gives indirect evidence in favor of the fact that the death rates are influenced by polluted environments. It must be taken into consideration, however, that these areas include the "oldest"-age population. This can also be said about the most challenging problem facing the population of Russia's Asian part: a short life expectancy (59.9 in males in 2002, with 7.2.5 of females).

The past decade has witnessed a growing social strain as a consequence of the ever worsening socio-economic status of population, an increase of death rates, too early deaths, and an escalating number of persons suffering from various forms of inborn pathologies. In Asian Russia, two polar types of territories can be identified in the distribution of social strain indices. One includes Siberia's large industrial areas, and the other encompasses the poorly developed territories in Siberia's north and east.

Many indices characterizing the life potential of Asian Russia's population, are considerably lower than the Russia-wide average. Compared to other countries, Russia in general and Siberia in particular occupy a peripheral position as regards the death rate and life expectancy. The specific natural and socio-economic conditions are represented by the death rate indices associated with fatal accidents, poisoning and injuries, and infectious and parasitic diseases which are higher for Asian Russia when compared with the country's average. In 1965, a tendency for an increase of death rates was observed for ages 30+. In 1982–1983, the highest death rates were recorded for 20 to 50-year-old males. Since 1990 the death rates of adults and children have been escalating. Combined with the ever decreasing birth rates, this has given rise to decline of total population (a so-called "depopulation" effect).

Sustaining and enhancing the life and labor-resource potential of the Siberian and Far-Eastern population is of paramount importance and constitutes a first-priority goal in optimizing the territorial organization of life activities.

It should be noted in conclusion that the unstable situation in the economy, combined with the uncertainty of the people, lead to an increasing number of one-child or childless families.

Demographic forecasts are also distressing – Russia's population, including its eastern regions, would be declining every year. Under the present socio-economic conditions, it is impossible, in the immediate future, to resolve the problem of providing for an increased reproduction of the population. It seems likely that the depopulation processes can only partially be decelerated by, for example, reducing infant's death rates through a bettering of health centers and a regulation of migration flows.

Considering the country's current economic and political conditions, it is important, at a regional setting, to plan a principal strategic line of development. It would help to elaborate a special attitude toward the eastern regions, and to comprehend the special geopolitical and geo-economic role played by the Far East and Siberia for the Russian Federation in general. At an early stage, the regional policy should be built around the state-support priorities as regards taxation, electric energy and transportation rates, and customs privileges.

The region urgently calls for an effective demographic policy because of the pending historically determined inconsistencies in the population distribution. This is because Siberia and the Far East have been and are regarded solely as a region supplying raw materials, which is aggravated by the controversies in the regional redistribution of the labor potential, a dramatic outflow of the population, a worsening of the reproduction processes, and by the unfavorable ecological situation observed in a number of regions.

Thus work was done with financial support from the Russian Humanities Research Foundation (05-02-9120a(Uk).

REFERENCES

1. Ryashchenko S.V., Misevich K.N., Chudnova V.I. The theory and meth odology of regional socio-ecological analysis in Asian Russia. The Tenth Scientific Conference of Geographers of Siberia and the Far East. Irkutsk, 1999.
2. Misevich K.N., Ryashchenko S.V. Socio-ecological problems of population activities in Asian Russia. Geografiya i prirodnye resursy, 1997, 3.
3. Vorobyev V.V., Misevich K.N., Vorobyev N.V. Socio-demographical problems in Siberia and the Far East. Geografiya i prirodnye resursy, 2001 2.
4. Vorobyev V.V. The Population of East Siberia: Present Dynamics and Prediction issues. Novosibirsk: Nauka, 2001.
5. Vorobyev N.V. Regional Organisation of Population Migration in Siberian Conditions. Novosibirsk: Nauka, 2001.
6. Ryashchenko S.V. Regional Anthropoecology in Siberia. Novosibirsk: SB RAS Publishers, 2000.

CHAPTER

13

A Geographical Understanding of The Ethnicity

M.V. RAGULINA

The current period at the turn of the Millennia is marked by a contradictory phenomenon: integration and globalization of technological, informational and social aspects of the life are proceeding hand in hand with the striving of mankind to conserve its ethnocultural multifaceted profile. The processes that have started in the second half of the 20th century (disintegration of the colonial system, modernization of the economy, build-up of the nuclear potential, and large-scale massive migration) involved a decay of the positivist approaches. A characteristic "intellectual crisis) that preceded the "turning point of the epochs", can be most thoroughly appreciated within the context of post-modernist views. A fragmentation of time, "loss of sense", intertextuality, nihilism, and historical amnesia are only a few of the basic characteristics of post-modernism philosophy. The products of this deconstructive work are highly valuable for ethnographic research, as the emphasis is for the first time placed on the recognition of the "otherness" of different cultures, and of the existence of a peculiar system of "double coding" that "can be decoded through literature, arts, and culture ... , in order to use this arch-essential definition of post-modernism as a strategy of communication with different levels of society" [1, p. 73].

A distinguished American scientist M. Fisher maintains that the prime objective of cultural anthropology as regards its applications is the activity that "ensures a maximally informed cultural translation in the numerous groups of the extremely differentiated today's society. The applications context must identify the framework around which to build mutual understanding where it is lacking ... the problem is not merely that people do not understand one another but that their "misunderstandings" are constantly and stably reproduced" [2, p. 7].

In the present conditions, apart from the ubiquitous intercultural and inter-group misunderstanding there arises a curious paradox implying the "lack of the desire" to understand in the milieu of scientists, who, as would seem, ought to investigate the essence of the phenomenon and "launch bridges". Thus, V.A. Tishkov, while advocating the fictitious nature of the "ethnic" and "social" categories, identifies, nevertheless, the dominant (hegemonist) and protective (peripheral) types of nationalism [3]. However, it is unlikely that the present Russian meaning of the ethno-national situation is consistent with such an approach.

Attempts made by some scientists to unravel the essence of ethnicity lead to stunning results: ethnicity is the pre-empirical and, moreover, non-empirical principle of organization of mankind, a characteristic "super-social substance" [4, p. 45], or, in its more pragmatic version, ethnicity is a peculiar social window" into the world, and a real, rather than illusionary, integrated totality of social relations between people [5, p.29]. S.V. Cheshko [4], while considering this phenomenon a variety of social identity, pointing out the elusive nature of the ethnic that "can manifest itself everywhere, and influence any spheres of human life and activity and, at the same time, it is nowhere present", poses an "accursed" (as he put it, in the sense of unsolvability) question: For what reasons do people elaborate "ethnic constructions"? (p. 38).

Indeed, to investigate the essence of ethnicity is an urgent, critical task of science. The unsolvability of the "accursed" question, however, is in many respects the result of an artificial narrowing of the theoretical basis of ethnology, its "pupation" and its being closed to the influx of philosophical ideas, which is also pointed by ethnologists themselves [6], let alone the interaction with geography. Furthermore, the last factor lays the groundwork, we believe, for ample, many-sided opportunities as regards a productive theoretical mutual enrichment. The ethnoecosystem approach which serves to represent (in the opinion of some authors) geography in ethnology, reflects only a small part of this contact.

Most ethnologists, who responded with a harsh criticism of the definition of the ethnos as a behavior stereotype in landscape that was suggested by L.N. Gumilev, were subsequently confronted by the "wall" which they had created and which was indeed insurmountable within such a context: "... What is the origin of the ethnic in general and what are the requirements and aspects of human activity that gave birth to it, and what is the function inherent in it only?" [4, p. 38].

The basic statement, suggested by B.M. Ishmuratov [7], which make sit possible to remove the existing contradictions and opens up the avenue for a further theoretical quest, implies that "the differentiation between social and ethnic organization leans ... upon distinguishing the patterns of development of these objects – social in one case, and geographical (biogeographical) in the other case. Consequently, these forms are both two structural levels of organization of the same phenomenon, mankind, which expresses the duality of its natures – biological and social at the same time" [p.7].

Within this context, we try now to give our answer to the questions raised, in the form of a hypothesis. It is based on ethnogeographical data, some concepts suggested by L.N. Gumilev, B.M. Ishmuratov's 'district' approach, as well as on published data on behavioral geography, communication theory, linguistics, and neurolinguistic programming.

The tightly intertwined duality of the biological and the social in the ethnos is manifested in its interrelations with its environment which, while satisfying the life-supporting (primarily material) needs, is converted in human minds to a sign system of symbols and images that has a powerful influence both on the spiritual self-realization of society and on nature management (the sum of technologies) combined into a common life-supporting complex.

According to the information theory of communications, Man constructs objects of his environs by attaching to them a particular meaning, based on his own activity [8]. Yi-Fu Tuan believes that one of the objectives of this research area is to enhance our understanding of "how the localities and the environment in general turn into pronounced feelings and thinking; moreover, the experienced (thinkable) quality of space can never be wholly understood through the description of natural (physical-geographical) structures, and no one of the human ways lead there" [9, p. 372]. Such a subjective aspect of the life-support process is experienced in the culture (ethnos) – territory linkages.

The adaptiveness of pre-ethnic cultures is axiomatic; on the other hand, the "nurturing landscape" of the Oikumena", even during the period of making of mankind, was not a stringent determinant of culture. The originality of closely lying archeological monuments dating from the early and middle Paleolithic Period, bears witness to the inception of cultural differences and, hence, of the beginning of the ethnic-formation process. Also, the habitat and ethnocultural traditions of the human population are in the process of mutual adapting and co-creativity, giving rise, as B.M. Ishmuratov [7] put it, to a regional-ethnic complex of the economy that provides for survival and reproduction of the community. The evolution of cultural, anthropological and linguistic characteristics proceeds at different rates. As time elapses, the rate of these processes increases; on the other hand, even for the primitive-communal system, archeology has a large number of examples of how rapidly ornaments of ceramics and implements of labor are changing in their form.

The linguistic factor of ethnic formation "introduced an additional coordinate into the array of genetic barriers that have basically a geographical ad biological character operating until separate languages are created" [10, p.48]. The languages of the antiquity were largely flexible, and the interaction through a separate contact produced their "affinity": extrapolating the data on Australian aboriginals to communities with a primitive appropriating economy, linguists arrive at the conclusion that any adult either was fluent in one or two dialects or understood them [11].

M. Highdegger's aphoristic idea that language is the home of existence, explains the interest in this realm that is remote (as would seem) from geography. Indeed, with the advent of language, man begins to manipulate with symbols thus creating (modeling) his world and himself in the world. By penetrating deep into modeling phenomena, it is possible to come very close to ethnogeographical interpretation of the cultural and cross-cultural phenomena.

Perception as a psychic process is limited by the capabilities of man's sensory channels: physical phenomena lying beyond them are simply not perceived. For instance, the hearing organ can perceive sound waves in the range of 20-20 000 oscillations per second, and the eye records the waves between 380 and 680 millimicron. The active process of perception in the modern area of psychology, neurolinguistic programming (NLP), is regarded as the first "system of filters", or neurophysiological restrictions, which is responsible for the distinction of the world from its human representation [12]. This system is the basis uniting all members of mankind as a single biological species.

The second, socio-genetic, "system of filters" imposes constraints on individuals as members of a group. Its function implies selecting information arriving from the environment, in accordance with the goal of adapting and survival of the community. The founders of NLP use, as a generally accepted filter, the language system, indicating the presence of other categories of the "most diverse functions, with regards to which a given society has reached a relative consensus" [13, p.124].

The language system reflects the character and depth of human experience. This is especially perceptible when looking into ethnic communities with the appropriating economy, the life-support system of which is comparable with the rhythms of the natural environment. Furthermore, the wealth of linguistic differences for the categories of experience is equivalent to the wealth of activity (choices) in this area and is associated primarily with survival in particular landscape-geographical conditions.

For instance, in the Evenki language "reindeer" is designated by several tens of words, while what is commonly named "snow" has more than 30 synonyms. However, such detailing exists only for vital spheres of contact with the natural environment: thus, in the language of South-American Indians 'hopi' is only one designation of flying objects except for birds (insects, aircraft, pilots) [14], while the language of the North-American maidu contains for the description of the entire color spectrum only three words, although the human eye is able to distinguish in the visible spectrum 750 000 varieties.

To put this another way, the world, remaining a multifaceted one, serves as a basis for constructing the model representing primarily categories which ensure survival and reproduction of a particular ethnic group under particular environmental conditions. This is most dramatically evident in the case of so-called "traditional cultures", which are based on ethnical nature management (ENM). However, such a regularity is also observed in peoples with their more "technologized" attitude to nature.

As a result of investigating the adaptation of the Russian old inhabitants in the Transcaucasia [15] using the semantic differential method, it was established that having lived in the new conditions for about 150 years and calling them the "native", they "have kept the unconscious psychological preference to the natural landscapes of the middle Russia which their ancestors had left in the 1830s-1840s. For the Russians, the main semantic characteristics of native nature, as contrasted with foreign nature, are: soundness, stability, and generosity; while for the Azerbaijans: lightness, mobility, and changeability" (p. 102).

Obviously, the sources for such an approach reside in the "initial" types of sedentary and nomadic management of nature. We believe that the ethnos, while living through the historical life span when it is interacting with its habitat on the basis of ENM and most fully displays its adaptive potential, "imprints" the main polyfunctional links in the spiritual tradition, as well as in the material tradition that is inseparable from it.

F. Ratzel [16] pointed out that "the main law of anthropogeography implies that things move together with people, through people, in their soul as the idea of the form ... this creates a close relationship between the people, their implements, weapons, adornments, etc. and attaché\s to them an ethnical character. Therefore their geographical distribution is indicative of the range of distribution of the people or, at least, of their links" (p. 399).

Inter-generation translation of "imprinted" experience is perceptible in all spheres of the ethnic existence. Subsequently, evolving toward "general technological" models of nature management, this adaptive potential of culture does not disappear without leaving a trace but joins in a more flexible and more complicated "system of filters" along with other factors, providing a given group the general in the perception of the world, which does render it an ethnical integrity. It seems likely that L.N. Gumilev [17] was right when he wrote that "although the ethnical belonging is indeed found in the consciousness of people, it is not the product of consciousness itself. Obviously, it reflects some aspect of the human nature, a much more profound in this case; thus self-consciousness is merely a distinguishing sign, rather than the essence of a subject" (p. 195).

Thus each ethnos generates its own model, a representation of the world. The length of this "designing" process, and the continuity of the "designer's elements" in almost every generation gave birth to a plethora of interpretations: from geographical determinism and behaviorism to the use of Young's psychoanalysis in geography. For instance, E. Maier, investigating the translation of the territoriality phenomenon, establishes the connections of the individual, nation and group of nations through the collective unconscious, where the main archetypes of human psyche. Such an approach is not indisputable, yet "archetypical" categories common to most peoples, such as images of the "world tree", vertical stratification of the Universe (the lower, middle and upper worlds), horizontal-radial schemes, and the organismic Cosmos that counteracts the Chaos, are embodied in their cosmological systems.

Cosmology is a peculiar "theoretical outcome" of modeling the world. The underlying ideas of the structure are maximally generalized: for instance, vertical stratification was observednot only in Shumers , Egyptians, medieval Chinese, and Pueblo Indians [19] but also in almost all Siberian peoples, and many others. Nevertheless, each of the cosmologies "ensures and operational and practically convenient system of orientation in space that is needed in everyday life" [14, p. 161] and is composed by the most important elements of the social and natural environments.

Thus, the Yenisei Kets' concept of the world is built according to vertical and horizontal vectors and reflects particular landscape-geographical habitat conditions of the ethnos. According to E.A. Alekseyenko [20] [20], the world of the "Kaduks" has lexical and semantic parallels (horned animal – mountains-circles) and contains analogies to a similar division in the Central Asian and South-Siberian cultural traditions. At the same time there exists a "bansel" (*ban* means earth), i.e. a "fluvial" horizontal system of the world which has analogies in the data on the Ob Ugrs, Selkupes, and Evenki. The Evenki believe that at the river outlet, at the piedmont of the upper world, there are worlds of unborn souls, downstream in the form of a large reindeer moss pasture, unborn calves [21].

M. Eliade, describing the perception of the microcosm in archaic communities, observes that "Beyond this confined, closed little world, the unknown, the formless begins. On this side, a correct space exists, because it has long been inhabited and is well organized, whereas outside of this known space, beyond its limits, there are unknown, hazardous regions dependent on demons, spirits, the dead, and foreigners – in other words, the world of chaos, death and night. This image of the inhabited microcosm has persisted in cosmological views even of such highly developed civilizations as Ancient China, the states of Mesopotamia and Ancient Egypt" [12, p. 168].

There is a difference between cosmology and the real life of the ethnos : the former tends to symmetry and completeness, postulating "how it ought to be", and the latter expresses the dynamism of "how it actually is" .

In terms of a modeling of the ethnical image of the world, the theme of mutual influence of the cultures looks somewhat differently. We suggest that the "inner space" of this image that includes a set of stable views of the world, its group identity, and of the best strategies of interactions with it, be called the "dividable reality" of the ethnos. According to ethnographical sources, even the smallest, geographically isolated groups have historical experience that is embedded in language, mythology, and folklore and allows the dividable realities of different ethnoses to find overlap regions. Furthermore, results of cross-cultural interactions are directly dependent on the observance of the integratedness principle. In other words, there is an alternative: to perceive the contacting sides as independent entities or to try to impose changes, based on one's own model, i.e. in accordance with one's own views of one else's wealth.

The former variant can be supported by an example of the Evenk-Russian-Buryat interactions at the period before collectivization. Smooth processes of mutual acculturation left their traces in the three respective languages, traditions of the economy, and in the shaping of symbiotic life-support systems. For the beginning of the 20th century, one can identify three main types of nature management for the Evenki.

1. Reindeer breeding and hunting that is characterized by mobile methods of transport reindeer-breeding, with a significant land area under development. The subject of nature management is a nomadic territorial-neighboring community, the regulator of the principal production and spiritual relations on a social-cultural level.

2. Stock raising and hunting that temporarily precedes the third type and is conditioned by the acculturation influence of the neighboring ethnoses: the Russians, the Buryats, and the Yakuts. The basis for this variant is provided by families having no reindeer or only a small number of these animals. The economic cycle is determined by the requirements of stock breeding and hunting and is associated with the long-lasting settled life and semi-settled life; the interests of reindeer-breeding are of secondary importance.

3. Agricultural-stockbreeding-hunting that implies settled residence in timber houses, and a seasonal hunting-based nomadic style of life of male hunters. It is the strongest expression of acculturation processes: the presence of fertilized meadland farming, arable lands, a kitchen-garden, and dairy cattle is an essential prerequisite to final switchover to the Russian and Buryat economic models. There are no reindeer in the farmsteads or their numbers are only minor.

In the second and third variants, the nomadic community transforms into a semi-settled and settled community and, as regards its functions, is becoming similar to a peasant community consisting of old residents.

By the beginning of organization on cooperative lines, the economy of the aboriginal taiga ethnoses was not of the same type but represented a continuum of regional and local models of nature management that had shaped themselves over a long period of development under the influence of various ecological, economic-geographical and ethno-social factors. Each of these models is distinguished by a specific character of natural resource utilization, the rhythms and dynamics of land development, and by intrinsic land turnover.

The second variant of development, the forcible breaking of ethnical nature management, can be illustrated using the example of the 1930s. The original concept of how to change the traditional economy was based on paternalistic and stadial approaches. Ethnical nature management was considered obsolete and archaic, and letting the aboriginal peoples "live on their own" was regarded as the analog to their becoming extinct; if the ethnic groups "switch over by themselves to the new forms of economic management…, they should be let do so" [23, p. 1].

To accomplish this, the following major measures were planned: 1) to create aboriginal hunting enterprises and allot area to them, which would be prohibited for the main population; 2) to set up cultural bases, or administrative, economic and cultural centers; 3) to gradually switch over to a settled way of life; 4) to transfer reindeer-breeding to a large-site sectoral model, and to a further development of livestock husbandry; and 5) to provide them through cooperation with the necessary commodities and services.

For political reasons, however, the actual development of events had a harsher character. Tofalaria was selected as a "testing ground" for a first shift-over of the nomadic ethnos to a settled way of life. Based on unique archival documents, it is possible to reconstruct the course of events. Collective farms that were organized, essentially were communes. "Everything was collectivized, from yurtas to sporting guns, and dogs. Collectivization ... was completed within one month, and large amounts of credits were granted to all those joining collective farms. The Tofalars had since seen no money at all, and all functions involving sales of fur-skins, and the supply of ammunition (in the form of obvious egalitarianism) were performed by the heads of collective farms, the Russians, as it was believed that the Tofalars could not cope by themselves with the tasks of managing" [24, p.30].

The entire system of ethnical nature management underwent drastic changes. Comprehensive development and tapping of life-supporting resources was intentionally substituted for by the setting up of "species-oriented" hunting teams, or, separately for sable, squirrel, and the ungulates, and specialized fishing teams; furthermore, women's teams were organized – for gathering berries and mushrooms. Socialist emulation was a key factor for achieving an annual growth of plan targets.

Collective reindeer-breeding was started with the creation of collective farm-owned herds: "hunters received reindeer foe a fee, and reindeer owners had to pay for their own reindeer" [24, p.31]. Shortly afterwards, in 1930, a draft law of reindeer-breeding was suggested and approved by the Legislative Assembly of the Evenki ('Suglan'), on the Communist Party District Committee's initiative: "For the purpose of instituting proper management and supervision of this sector, as well as of carrying out in a planned manner the necessary measures ... the merging of reindeer into collective farm-owned herds shall be recognized as insufficient ... it is becoming necessary to amalgamate reindeer-breeding, an independent sector of the Tofalar economy at an eventually large scale" [Ibid.]. As a result, reindeer-breeding was at the verge of total elimination. The many-century economic experience was in sharp contradiction with innovations. As the Tofalars, like the taiga Evenki, had never pursued any pure, "sectoral" hunting or reindeer-breeding, the "dissolution" of these kinds of nature management that merged into a common system, was in conflict with the population's labor habits.

The former, harmonized, social links were also decaying. The greatest destructive influence was exerted by the use of collective structures for the repressive purposes, and the aftermaths of those measures are still experienced to date. In accordance with the policy of liquidating the kulaks as a class, in 1932 it was decided by the Aboriginal Soviet that the "kulaks" (already devoid of the suffrage), who included those owing more than 20 reindeer, and shamans (mostly the poor) were imposed sizable penalties and incurred forced labor [24, p.51]. The document about exaggerations in collective farm building pointed out: "The kulaks and shamans were assigned unrealizable firm tasks, followed by the confiscation of all property, even teapots and sacks, and they were fined and evicted beyond the district" [24, p.30]. Those repressive measures were carried out before the former commune members' eyes (and with their participation!), who had been maintaining neighbor, kinship and religious ties.

In the ethnical cultures of the taiga hunters and reindeer-breeders, whose life support system is based on unstable sources of subsistence, human life was always regarded as the supreme value, while the traditions of mutual aid had been a crucial adaptive mechanism of survival under harsh environmental conditions. Breaking of the norms and habits that had been observed for centuries, involves not only a decay of system-forming ties of a community but also a destruction of an individual's previously integral concept of the world. Thus the infringement of the integratedness principle inevitably causes opposition and a peculiar pathological adaptation to the "civilizing endeavors of the 'Kulturtraegers' (persons with civilizing mission)".

The present set of challenging problems facing the small ethnoses is sharply pronounced not only in the sphere of the material life support system: social-psychological apathy, and alcoholism are markers of destructive changes in the model of the world. For instance, medical scientists, who investigated the ethnical characteristics of alcoholic psychosis in the Nenets, point out its "occupational trend": hunting scenes, setting fishing nets in the homestead, and manipulation with knives. Thus, "once the state of psychosis is over, there is almost no critical reappraisal of the event. Patients cannot find any depressive nuances of their state of mood, such as the sense of fear of their health, a burden of guilt before their neighbors and relatives, or those psychological feelings which are common in a drug-abuse clinic and contribute a great deal to a further active anti-alcoholic treatment" [25, p.89].

This is a typical situation: destructive behavioral stereotypes are reproduced, and only a tiny portion of cultural heritage is used. To get over a crisis, as demonstrated by recent experience, mere independence would not suffice – the gaps in the inter-generation ethnical tradition requires aid from specialists, at least during an early stage of ENM recovery. Its goal is to close information lacunas. While joining in the model of the world inherent in a given ethnos and respecting its integrity, efforts should be made to attract those resources which it has already.

To conclude, it should be remarked that the rules and traditions used in developing by a group any space are determined by the ethnical model of thee world that has been firmly established on the basis of a selective perception of the reality, for the purposes of achieving the best life-support system and spiritual self-realization of the people. Furthermore, the more adaptive is their economy and the less developed is the "general technological" base of production, the greater the extent to which a given community needs real interrelations with their habitat, ethnical nature management.

It is not only the economic requirement – there are a lot of sad examples of how many countries tried to save, by means of paternalism, "disappearing" aboriginal cultures, by attaching them the role of living museum exhibits able to make souvenirs. The chief function of ENM is that of conserving the quality of ethnicity itself (or, as we understand it, the ability to reproduce itself and the world in terms of a unique (inherent only in this people) model of the world and, on its basis, to adequately interact with its natural and social environment). ENM and the closely associated cultural traditions ensure inter-generation continuity of the ethnicity phenomenon of so-called small northern peoples.

At the ENM, ethnicity as a fundamental characteristic of a modeling of the image of the world is clearly pronounced and evident in all spheres of material and spiritual culture. With the progression of the productive forces, a growth of the population, an accumulation of the information and technological bases, there is taking place an inevitable bias from ENM toward more economically efficient "general technical" nature management. In this case, ethnicity is translated largely in the social-domestic, cultural and psychological spheres to become, as paradoxical as it would seem, more stable to impacts of the territorial environment and to acquire the very "elusiveness" according to S.V. Cheshko.

A continuum of the models of the world for different ethnoses, each of which reflects some unique aspect of life, constitute a kaleidoscopic picture of the ethnical being of mankind.

This work is done with support under RF President grant MD-5510.2006.5.

References

1. Folsh-Serra M. Geography and post-modernism: Linking humanism and development studies. The Canad. Geogr. 1989, vol.33, No. 1.
2. Reflections on the fates of science. Etnograficheskoye obozreniye, 1996, No. 6.
3. Tishkov V.A. To forget about the nation (post-nationalist understanding of nationalism). Voprosy filosofii, 1998, No. 9.
4. Cheshko S.V. Man and ethnicity. Etnograficheskoye obozreniye, 1994, No. 6.
5. Zarinov I.Yu. The historical framework of the ethnicity phenomenon. Ibid., 1997, No. 3.
6. Reflections on the fates of science. Etnograficheskoye obozreniye, 1996, No. 5.
7. Ishmuratov B.M. Economic-geographical and geopolitical factors of development of the inter-ethnic relations at the present stage. Geografiya i prirodnyte resursy, 1997, No. 2.
8. Current Foreign Social Psychology. Texts. Moscow, 1984.
9. Yi-Fu Tuan. Sign and metaphor. Ann. of the Association of Amer. Geogr., 1978, vol.68, No. 3.
10. Alekseyev V.P. On the Earliest Stage of Ethnic Formation. The Ethnos in Pre-class and Early-class Society. Moscow: Nauka, 1982.

11. The History of primitive society. The Epoch of the Primitive Clan Community. Moscow: Nauka, 1986.
12. Bendler R., Grinder D. The Structure of Magic. Neurolinguistic Programming. SPb.: Bely Krolik, 1996.
13. Personality Transformation: Neurtolinguistic Programming. Odessa: Khadzhibei, 1995.
14. Gold D. Psychology and Geography. The Fundamentals of Behavioral geography. Moscow: Progress, 1990.
15. Lebedeva N.M. Psychological aspects of ethnic ecology. Etnicheskaya ekologiya. Moscow: Nauka, 1991.
16. Ratzel F. Anthropogeographie. Stuttgart, 1891, vol. 2.
17. Gumilev L.N. Ethnos and landscape. Izv. VGO, 1968, vol. 100, No. 3.
18. Maier E. Torah as moveable territory. Ann. of the Association of Amer. Geogr., 1973, vol. 63, No. 4.
19. Yi-Fu Tuan. Ambuiguity in attitudes toward environment. Ann. of the Association of Amer. Geogr., 1973, vol. 63, No. 4.
20. Alekseyenko E.A. Ethnocultural paralleles of the study of shamanism in the Kets. Ethno cultural Contacts of the Peoples of Siberia. Moscow: Nauka, 1984.
21. Vasilevich G.M. The Ewenki. Historical-Ethnographical Essays. Leningrad: Nauka, 1969.
22. Eliade M. Images and Symbols. Moscow: Nauka, 1993.
23. Petri B.E. The Cultural center Project for Siberia's Ethnic Groups. Irkutsk, 1928.
24. Survey Data on the Tofalar Indigenous Council for the Year 1933. GAIO, f.r. 2375. Op.11, d.97, pp.29-32.
25. Sidorov P.I., Shubin S.V. Ethnic characteristics features of alcoholic psychosis in the Nenets. Sotsic., 1994, No. 7.

CHAPTER

14

The Pivotal Role of the Baikal Region in the Development of Siberia

B.M. ISHMURATOV AND K.N. MISEVICH

At the turn of the century, the socio-economic situation showed some features of a long-expected stabilization, namely, an enhancement of the society's attention to the issues of forecasting and strategic planning, and the search for programs and avenues for the country's most efficient further development for the several decades to come. Of particular importance in this regard is the issue of assessing the potential of Russia's different regions, including its Asian part, in the overall progress.

A task of the development strategy can involve achieving such a place in the economy, science, technology, and culture that would ensure Russia's stable position among the most advanced and influential states of the globalization era, as well as its sovereignty and territorial integrity plus a maximum possible satisfaction of the material and spiritual needs of its citizens. Elaborating such a strategy is equivalent to a critical review of all achievements and negligence and omissions inherited from the preceding stages of development.

Any Russia-wide program of social, economic, political or cultural development cannot be considered complete until Siberia's potential and importance is properly taken into consideration. Economic development of its regions has led to highly erratic results. The country's planned industrialization that has its origins in the years of the first five-year plans, was beneficial to an increase in coal production, the construction of large metallurgical (iron and steel) integrated plants and works, the emergence of large cities, and the electrification of industries and public utilities. In the years of the Great patriotic War, tens of plants and factories were transferred to Siberia from European regions. The post-war years saw a large-scale construction of hydropower plants on the Angara, and then on the Ob and Yenisei; there emerged new machine-building plants, integrated logging-lumbering enterprises, plants specializing in processing chemical raw materials, various industries, and tens of establishments of higher education.

During the so-called "Great Construction Projects of Communism" that culminated in the construction of the Baikal-Amur Trunk railroad, citizens were resettling to Siberia from Western areas; this is especially true for the period of development of new virgin and idle lands. Of paramount significance for the country's economy was the project of development of the West-Siberian oil and gas bearing field that started in the 1960s. Exceptionally high is the share of Siberia in the production of petroleum, gas, coal, and electric energy, in the timber industry, and in the production of gold, diamonds, and furs.

On the other hand, in the economy of the Siberian regions there exist large disproportions and shortages. Most of first order goods are imported from the European part of Russia or from other countries. The level of living in Siberia is much lower when compared to the other regions; the expenses incurred are also significantly higher because of the severe climatic conditions, long distances, and expensive staple foods. For that reason, the population has long been reducing, especially in northern regions. A sharp increase in transportation and electric energy tariffs, combined with low competitiveness of Siberian products, led to a significant decline of production, both in the industry and in agriculture. The richest natural-resource potential here is not utilized because of the lack of investment, infrastructural and (not infrequently) personnel possibilities. As a result, Siberian inhabitants, and also the country's higher authorities are discontented: the consistently predicted "golden mountains" won't become a reality. The recently expanding possibilities of economic exchanges with Asian-Pacific countries are also exploited inadequately.

It should be recognized that Soviet Power, at some period, was unable to correctly rank the tasks to be accomplished. The policy of accelerated development of the national peripheries that was declared in the first post-war years, has inflicted the greatest damage to Siberia, and the republics of Southern and Central Asia were unable to achieve the targets. If, however, we take into consideration the scope of scientific and technical assistance rendered by this country to China, North Korea, Vietnam, India, Cuba, and to many other countries, we find that Siberia can, indeed, produce many things at a large scale, holding much promise. Overall, the demographic potential of the initial (native) Russia, in the realization of imperial ambitions in the Baltic states, Poland, the Ukraine, on the Caucasus, and in Central Asia, was not sufficient for the assimilation of the periphery or the development of Siberia which shared the fate of those abandoned Russian stick-in-the-mud areas.

Development of Siberia, without any special-purpose long-term strategy, has become a trap for Russia, and the myth about the inexhaustible or exceptionally valuable Siberian wealth was merely a myth if we compare the benefits with expenditures. In the greater part of Siberia and the Far East, annual growth of biological output is negligible, and biological products need hundreds or thousands of years to accumulate. Under these circumstances, a concentration of the population is virtually impossible; for that reason, Siberia saw the advent of numerous peoples with a settled way of life. Nomadic gathering of biological products of the taiga and tundra turned out to be the only possible form of utilization of this wealth. It is beyond reason to hope for any radical change of the situation in the foreseeable future.

The chief drawback of economic development of Siberia and the Far East involved the inability to correctly set the territorial priorities, the neglect of the rule implying that efforts should be concentrated where they can be of the greatest benefit, and "smearing" or "spreading" over its giant expanses.

STRATEGIC SIGNIFICANCE OF THE BAIKAL REGION

Russia is a large country and, only for that reason, it cannot afford the building of its future by developing some single center, even a regional one. Most countries are characterized by a polycentrism, a drift of growth centers according to time and circumstances involved,, and by a constant enlargement of the number of centers of specific and, hence, advantageous development. This is particularly evident in the case of the country's geopolitical centers.

The public-political and economic development of Russia has been based recently on the northwestern (St. Petersburg), Novorossiisk-Krasnodar and Vladivostok centers, or "historical platforms". Common to them (with a strong difference in a great number of indices) is their bordering and, at the same time, seaside position. Even with the Capital's region, the aforementioned "pivots" are clearly too few for Russia. On the other hand, there are strong arguments in favor of the development of this inland, yet bordering region as a new historical platform, or a support base for Russia's strategy in its central part, the Baikal region.

The Baikal region in this case, unlike the aforementioned support bases of Russia, could (and must!) be evolving within a geographically uniform territory that is unified by its belonging to the watershed basin of Lake Baikal, a natural, cultural and economic phenomenon of global significance. According to the belonging to the Baikal watershed basin of at least a portion of the territory, this region should conceptually incorporate the Irkutsk and Chita regions, Buryatia, and the Republic of Mongolia (Fig.).

Macro-geographically, the territory under consideration has all attributes of integrity, economic-cultural and landscape-ecological homogeneity and common centrality (middle position) within Asia. Without going into details, special advantages of the region's geographical location include the presence of the mainland's largest rivers (Lena, Angara, Yenisei, Amur, and others), the belonging to a global watershed area, the interpenetration of economic and cultural structures of world civilizations (industrial-Christian, Buddhist nomadic, and taiga pagan).

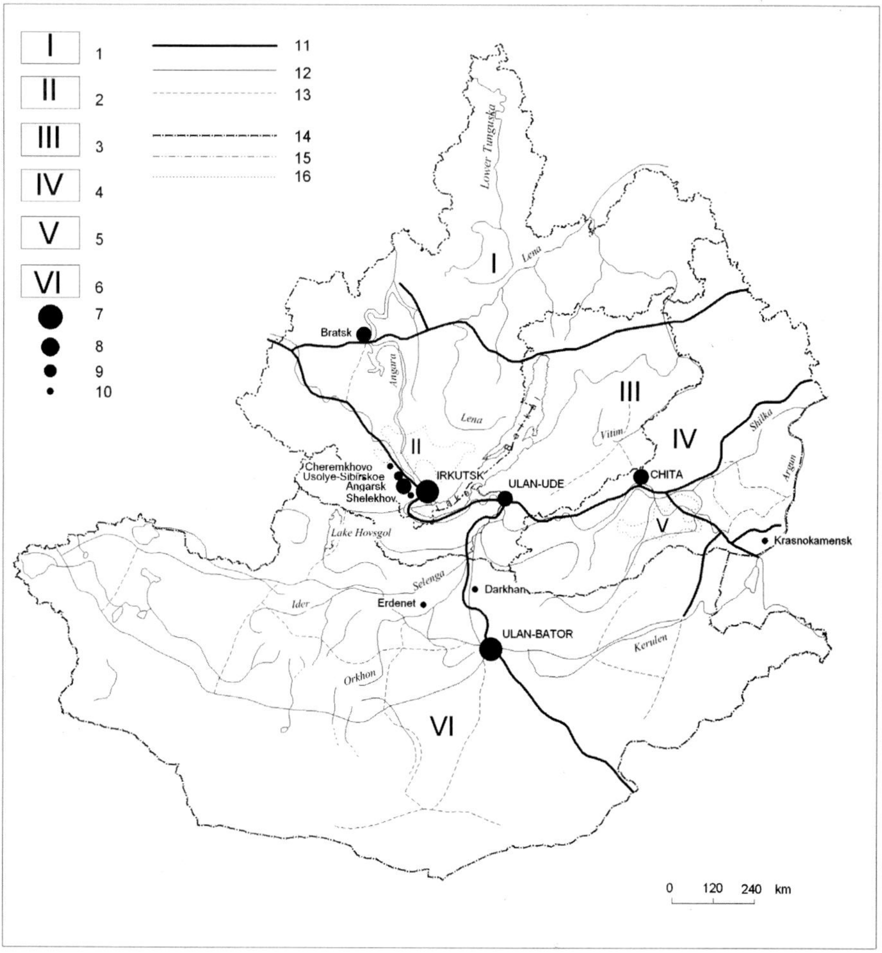

Figure: Administrative-territorial subjects forming part of the Baikal region.

Areas: 1 – Irkutsk region; 2 – Ust-Ordynsky Buryat Autonomous District; 3 – Republic of Buryatia; 4 – Chita region; 5 – Aginsky Buryat Autonomous District; 6 – Mongolia. Urban population: 7 – over 500 thousand; 8 – 200-500; 9 – 100-200; 10 – less than 100 thousand. Roads: 11 – railroads; 12 – highways; 13 – dirt roads. Boundaries: 14 – state frontier; 15 – regional and republican; 16 – autonomous districts.

The Baikal region constitutes a link of the trans-continental transit trunk railroads: the Trans-Siberian and Amur railroads, with their legs to Southeastern Asian countries (the former KVZhD (Russian abbrev. Meaning "Chinese Eastern Railroad") – from Chita through Borzya to Harbin, and from Irkutsk through Ulan-Ude to Beijing). With such advantages, the region may well become the principal link of accelerated growth of the productive forces, promotion of positive and desirable shifts of the geopolitical and socio-economic situation in Siberia, the neighboring Mongolia, and in the bordering territories, the development of which meets to the greatest extent the interests of Russia, and not only in the early 21st century.

Special mention should be made of the main functions of the Baikal region in the public territorial division of labor. It is the largest eastern industrial-economic complex; further to the east and north are only separate, mostly specialized industrial centers; it is the staring point and the base for development of Russia's Far-eastern and northern regions. The Baikal region is a proving ground for a number of nation-wide programs: energy generation, transport, gold and currency, and furs; production of non-ferrous metals, forestry, and some sectors of the defense complex. It refers to the main sources by which to meet Russia's internal and export needs for natural resources: water, timber, gold, mica, furs, and various ores and salts; it is rich in reserves of petroleum, gas, hydropower, etc.

Finally, the region performs ecological functions with global implications. It includes the world's main reserves of pristine clear freshwater, the center of reproduction of biological structures and systems lying in the Earth's moderate latitudes. Of significant importance is its role in the reproduction and functioning of taiga, steppe and desert landscapes of the Central Asian type. Recently such an abundance of water bodies, and forest and mountain landscapes of a spontaneous state is becoming an attractive site for an ever increasing number of recreationists from many countries worldwide.

The economic-industrial potential of the Baikal region, notwithstanding its breadth, needs revival as well as additions. Primary attention should be paid to machine building, especially the production of staple commodities (in the Chita region there are prospects for the erection of an assembly plant through cooperation with companies from South Korea); it is desirable to create enterprises specializing in the production of low-power facilities such as mobile hydropower stations, wind motors, electric motors and generators of various purposes, agricultural equipment, etc. The region provides the eastern Russia's only base for the production of building materials – in high demand by northern, eastern and southern neighboring regions.

The institutional and infrastructural groundwork for organizational, cultural-informational and educational functions of the Baikal region call for a serious further development, both with Russian and international implications. The development of the region as a new historical platform or a support base of

Russia in its eastern part is not possible until its cultural and scientific-educational influence on neighboring territories is enhanced.

In this context, one can speak of the creation of universities with a religious-culturological profile, such as the Orthodox-Christian and Buddhist-Confucian Universities, the Institute of Cinematography, centers for folklore and applied arts, the Academy of Asian Culturology, institutes of a humanities and basic research within the Russian Academy of Sciences and the Russian Academy of Medical Sciences, as well as the National Academy of Mongolia. The establishments of higher education and special education should maintain a regular exchange of students and teachers with Asian countries. The regional mass media need improvement and enhancement, both television and radio broadcasting, and publishing. Of course, the region is in need of its own conservatoires, new theaters, libraries, and sports centers, including those featuring world standards and significance.

DEEPENING OF COOPERATION WITH MONGOLIA

The people of Mongolia, as before, show friendliness toward Russia and hope for our assistance. We ought to immediately get rid of our indifference and, perhaps, miserliness in respect of Mongolia: the geopolitical benefits are evident if we keep our positions as a leading foreign partner in markets and in the economy of Mongolia, as well as the natural (based on the prevailing environmental regimes) attraction of the Mongolian economic systems towards neighboring southern Siberian systems; some commonalty of the ethno-historical structures and traditions justify an enhancement of the interaction and even mutual dependence of the Russian-Mongolian political guidelines.

The fixed assets of the Mongolian industry and agriculture have long not been brought up to date or modernized; products of these industries is confined within their boundaries, and the power generation base of the economy calls for a strengthening. Russia, as was announced in the Ulan Bator Declaration of the Presidents of Russia and Mongolia, is willing to render assistance with a modernization of the largest enterprises of Mongolia: GOK (the ore concentration integrated plant at Erdenet), the Ulan Bator Railroad, and other enterprises. A primary role in fostering the economy of Mongolia should be attached to the Baikal region. Some of the directions of such an assistance are already obvious: the construction of main gas pipelines and power transmission lines for exporting gas from the Kovykta gas field and electric energy generated by the Angara cascade of hydropower stations to countries of Southeastern Asia, participation in the modernization of the industry of building materials, rehabilitation of stockbreeding and revival of agriculture, etc.

A great deal must be done in order to provide national culture with skilled personnel, revive the traditions of teaching the Russian language in Mongolian schools. Training of the Mongolian young students in state-of-the-art technical specialties (repair and servicing of motor cars, domestic electric appliances, servicing of communal amenities and networks, etc.). In the future, it would be of significant importance to enhance geologic exploration surveys, and any help with the road and transport construction, the housing construction, the resolution of territorial-production problems, including such large-scale projects as the optimization of development of the city of Ulan Bator, protection of Lake Hovsgol, the creation of suburban agro-industrial systems, etc.

It would be appropriate to formulate (to a first approximation at least) the principles of the best exploitation of the Baikal region's geopolitical potential: 1) the creation of the Baikal region is declared by a legislative act of the Russian Federation and approved by the President; the participation of Mongolia in the programs of the Baikal region is legalized by decisions of competent authorities of the Republic of Mongolia, and by a special agreement between the Governments of the RF and Mongolia; 2) the Russian subjects of the region conserve their respective constitutional status subject to any changes in the course of global reforming of the administrative-political structure of the RF in the years to come; 3) organizational-managerial decisions at a regional level shall be taken provided all its subjects fully agree to; 4) the execution of the agreed development programs of the region shall be under the directorship of its Council represented by the Administration Heads of the subjects, chaired alternatively by one of them; 5) the activity of the specially created working bodies of the Council shall be carried out at Irkutsk; 6) coordination of the activities of all structures in the Baikal region with competent organs of Mongolia shall be carried out according to special plans; 7) extensive, open discussions of forthcoming issues related to the development of the region shall be carried out at joint sessions of legislative (representative) bodies of all administrative entities; 8) current issues related to the implementation of the development strategy of the Baikal region could be taken up by a body similar to the Council for Mutual Economic Assistance or the Economic Councils during the late 1950s to the early 1960s; and 9) funding of the development programs for the Baikal region implies setting up a special targeted budget fund at the RF Government, and a fund of joint payments of the region's subjects, on a long-term planned basis.

MAJOR PATTERNS OF DEMOGRAPHIC DEELOPMENT

The Baikal region (its Russian part) occupies an area of 1551 thou sq. km, and its population is 5074 thousand (as of 01.01.99). The demographic potential is obviously insufficient for tapping its natural wealth and for creating here a powerful, reliable 'outpost' in order to strengthen Russia's boundaries on the east. The region's population has long been closely connected with economic development. Its distribution during separate historical stages reflects rather clearly the specific character and predominant utilization of a particular natural resource or of a complicated combination of several resources – according to human needs and requirements, the interests of business, and the public policy.

The most important shifts in the development of productive forces occurred in the post-war years thus giving impetus to a significant growth of population. From 1959 to 1992, the population increased by 1525 thousand (or from 3686 to 5211 thousand), or by 29.3%. An annual population growth for the region averaged 46 thousand.

This process was accompanied by an increase of the number of urban centers, and by a reduction of the number of rural settlements. Thus, from 1959 to 1992, the urban population for the entire region increased by 1686 thousand, whereas the rural population decreased by 161 thousand. Urban centers with the overwhelming population, 71%, became centers where the economic and human potential was concentrated. They were characterized by a consistent growth of the number of residents, the various industries, a more advanced (compared with rural areas) social and transport infrastructure, and by a predominance of housing with services and amenities.

The urbanization process was proceeding at a high rate and was accompanied by the emergence of new towns and urban-type villages. While in 1959 there were the region only 29 cities and towns and 91 urban-type settlements, the respective figures for the year 1999 were 38, and 133. Large cities were gradually sprawling. Thus, in the Irkutsk region there emerged the Irkutsk-Cheremkhovo agglomeration combining the cities of Irkutsk, Angarsk, Shelekhov, Usolye-Sibirskoye and Cheremkhovo separated by distances of 20 to 40 km. At the turn of the century, rkutsk became a political, economic and cultural center of the new industrial cities.

The population was increasing through a natural growth as well as immigration. This process continued till the year 1992 when the population in the Baikal region peaked. After that, the demographic situation in the Baikal region underwent abrupt changes – in connection with a change in the country's political and economic situation. During 1992-1999 the region's population decreased by 137 thousand, with the Russians constituting the greater proportion, whereas the native peoples still show a slight growth. The proportion of migrants, both internal and external, is increasing. For instance, according to the expert data for the year 1999, more than two million Chinese citizens settled down in Russia, the overwhelming majority of whom settled in the country's eastern part. Over the last decade, the Baikal region saw the emergence of an unfavorable migration situation. The direction of migrants changed sign, with a negative balance of migrations.

Summarizing what has been said above, one can state that in the late 20th century the Baikal region turned into a territory with an ever reducing demographic mass. The processes of settlement and economic development changed sign. The long period of population buildup through a natural growth and immigration gave way to reduced reproduction and emigration. During 1960-1985 the region saw a gradual decline of birth rates – from 27.7 to 21.3 per 1000 inhabitants. From 1985 to 1998, it decreased by almost a factor of two (from 21.3 to 11.1). For the same period the death rate increased from 9.5 to 11.7 per

1000 persons. The premature death rate of the population of employable age (especially males) is high. Infant mortality, as before, is high – about 19.0 per 1000 newborn. Total indices of birth and death rates have had a serious influence on the rates of natural growth which decreased from 11.8 to -0.9 per 1000 persons for the period 1985-1998.

The population decline caused the ratio of age groups to change. While in 1989 the population younger than employable age was 1556.2 thousand (29.7 % of the total population), of employable age 2997.8 (57.2%), and older age 689.4 (13.1%), the figures for the year 1999, respectively, were 1232.7 (24.3%), 3025.2 (59.8%), and 804 (15.9%). For the period 1989-1999, the number of children under 16 in the region decreased by 323.5 thousand. There is a decrease in life expectancy (56.8 and 69.6 years in males and females, respectively).

Economic instability, social problems, and the decreasing life quality cause an increase of the number of single-child and childless families, with the result of a decrease in the average family size – only 3.28 persons in the region, and about 3.75 in the Baikal region.

Demographic forecasts of Russia's population, including its eastern regions, are not comforting. Calculations show that until urgent measures on economic stabilization and upturn, an improvement of social conditions and of the standard of living are taken, the population will be decreasing as before. By the end of 2015, the Russian population will be 134 million, i.e. the country's population will dwindle by 11.5 million, in Siberia and in the Far East by 2.2 million, and in the Baikal region by 350 thousand.

In addition to a reduction of the total population, its quality is also decreasing – physical health, and ethical and psychic status. Society is facing degradation and ethical disintegration. That the nation is ill is attested by a great deal of evidence. For instance, conscription where only 18-39% of men called up for military service are recognized as fit for military service by medical commissions. The health state of girls and young mothers and, as a consequence, of the newborn is also distressing. Mothers more frequently abandon their newborn in hospitals (about 1 or two newborn per 100). Such children overcrowd children's homes, with increasing numbers of street urchins. It is urgently necessary to pool together efforts of parents, schools, mass media, law-enforcement agencies, and the public at large in order to eradicate such unhealthy phenomena.

Under the present socio-economic circumstances, in the near future it is not possible to resolve the challenging problem to enhance population reproduction. The only thing that could be achieved seems to be a partial easing of the depopulation processes through, among other things, a reduction of infant death rates by improving the operation of public health institutions as well as a regulation of migration flows. There is no question that country's demographic situation (including its eastern regions) is deplorable, and one should not be under an illusion that it would be easy to rectify it. In any case Russia has to independently resolve the relevant problems and interact with all countries of the world. Society will have to adapt itself to the new demographic challenges as well as to look for economic and social answers to them.

To successfully address the challenges posed by the socio-economic and demographic development of the Baikal region requires serious interpretation and in-depth analysis of the on-going processes. There is a need for a clearly defined long-term and, at the same time, flexible socio-economic, demographic and ecological policy leaning upon the new approaches and taking into consideration the realistic potential of the Russian State's core interests.

As far as the demographic situation in Mongolia is concerned, it is more favorable when compared with the Baikal region, although Mongolia is one of the sparsely populated countries of Asia. Its area is 1566 thou sq. km, and its population is 2617 thousand, with the population density averaging 1.7 per sq. km. This country occupies the 120th place in the world as regards its population, and the 29th place among Asian countries, yet it is high on the list among 30 countries of the world as regards the growth rates of population. The population is distributed extremely unevenly, which is determined by historical and natural-climatic conditions as well as by the extent to which the territory is developed. The northern part of the middle and western portion of the country is most densely populated; however, the population density does not exceed two persons per sq. km. Even in the Central Aimak that includes the city of Ulan Bator with its population of 812 thousand, its mean density is only slightly more than 6 persons per sq. km. The least density (less than 0.5 person per sw. km) is observed in the Southern, Middle and eastern-Gobi Aimaks.

The population of Mongolia grows solely through natural growth. Migration provides only a minor outflow of population; this is particularly true for young people leaving for to Russia, the USA, Korea, Germany, and to other countries in searching for training, employment or trading opportunities. The country is characterized by a high birth rate, a relatively low death rate, a high level of natural population growth, and the large proportion of children as well as by families having six children or more, which has a positive impact on population growth rates. Thus, for the period 1963-1999 the country's population increased more than twice (from 1017 to 2617.4 thousand). The country retains a high natural growth: 26.0 per 1000, although, when compared with the year 1983, these figures decreased by 10.2 (from 36.2 to 26.0 per 1000 inhabitants). The population growth is accompanied by changes of the urban and rural residents ratio. During the last 50 years, Mongolia saw the emergence of new industrial centers, and new large towns and settlements. The largest of them are Ulan Bator (812 thousand inhabitants), Darkhan (75 thousand), and Erdenet (45 thousand).

Overall, Mongolia is notable for a favorable sex-age structure of population, with young people predominating, which sets a firm stage for enhanced population reproduction and the shaping of labor resources. Demographic forecasts show that the population of Mongolia will be 4.1 million by the year 2025.

THE POLITICAL SIGNIFICANCE OF SELECTING THE BEST DEVELOPMNT STRATEGY OF RUSSIA IN THE 21TH CENTURY

A vigorous, energetic, accelerated recovery of Russia's might and economic upturn of Southern Siberia is the core objective of Russia for the 10-15 years to come. Procrastination is the thief of time in this case – there is no other alternative at all. To resolve this problem the whole of Siberia cannot be coped with – it is necessary to strengthen, within the first decades of the 21st century, outposts, support bases and, more importantly, the most favorable (both climatically and geopolitically) southern regions.

Siberia occupies 1/10 of land globally, and it is not only an integral part of Russia but also the object of global human interests, it is an ecological-geographic, resource-economic and cultural-historical phenomenon of international significance. Siberia has long had the role of the main (and unique as regards some other items) source of Russia's key resources. And this role in the creation of the country's industrial and economic potential is unprecedented – if there were no Siberia, the Russian State would have to follow a different path. Of particular great significance for the future of our Motherland is the fact that some of Siberia's resources will, perhaps, play the key role in the destiny of all mankind. This implies primarily such resources of the 21st century as freshwater, air, and the living space.

The ongoing growth of the population on the globe, is accompanied by an ever increasing overcrowding in the main regions of its distribution, and by an escalating shortage of water and foodstuffs, mass diseases, and social strain. Under these circumstances, a redistribution of population would be inevitable, which has already been legalized by the UNO in the form of the right of freedom to select the place of residence; there will be an increasing pressure (ethical and physical) on holders of extensively and inadequately used territories, and "ecological" revision of the world is possible. It is no accident that some countries are making claims of the vast regions of the Asian Russia and the former Soviet republics.

Of course, there is no point in appealing to build a kind of the "Chinese Wall" in order to cut the country off from the external world and its problems. On the contrary, it is advisable to pursue an active trans-frontier policy of joint utilization of natural and ecological resources and conditions. At this stage, however, we cannot afford to remain the poorest, badly organized and unwisely developed part of our state. Without Siberia, Russia has no national future – this must be clear to everybody; therefore, we are insisting on implementing a targeted program "Siberia – Russia" rather than merely propagandizing philanthropically or doing only "paperwork".

Otherwise confining the role of Russia only to the exploitation of the energy and oil-and-gas resources of Siberia and the Far East for the benefit of our neighbors, servicing of economic transportation of Asian-Pacific and European countries, and to recruiting manpower from countries of southern and southeastern countries for tapping our natural resources to be exported (in spite of the usefulness of such steps) will only consolidate the ever enhancing role of the raw-materials appendage.

In the event of accelerating the development of the southern part of Siberia and its core, the Baikal region, the initial measures relating to investments, credits, and modernization of production facilities should be taken by the state. Strict centralized planning of such measures and the distribution of the required investment resources is, of course, not the best method of economic organization; yet, without repeating the former mistakes inherent in the administrative-equalizing approach, it is advisable to pursue such planning in order to achieve a "boom" in the key spheres by combing it with resources of the private sector and concession-based involvement of international capital.

In the leading sectors of the industry and scientific and technological services, small and medium-size business should be provided with all kinds of support. In the Baikal region, it is necessary to use differential rent of exploitation of the hydropower resources, gas, chemical raw material, and timber – primarily for a further development of urban centers, and render all possible support to the research-educational and cultural-information complexes that are of international significance.

CHAPTER

15

Geographical Problems of Management at A Local Level

YU.P. MIKHAILOV, N.M. SYSOEVA, V.P. MOSUNOV, T.I. ZABORTSEVA AND S.V. DUDENKO

The so-called 'transitional period' towards creating a new socio-economic environment in Russia has posed a multitude of problems, upon the solution of which the current and future condition of society and the State depends. Among them are the key issues related to governing the life of society, and above all, government at a local level. A local level is characterized by the fact that it is precisely here that all public and regional policies of economic and social development should be implemented.

The problem of raising the standards of living in the country, which is the ultimate aim of reforms, has acquired a clearly pronounced territorial character. A characteristic feature of the former 'Planned Economy' was its normative approach in determining the quality of life of the population, and the striving to smooth out the contrasts among different regions and different types of settlements. The liberalization of economic relations led to a restructuring of the driving forces of local socio-economic entities.. A sharp differentiation in the levels of territorial development, and a reduction in the reproduction potential of many of them (agricultural and old-industrial regions in particular) is taking place.

The rural areas continue to degrade and remain in a state of deep depression. Every year, 3-4 thousand rural settlements disappear from geographical maps of Russia. Previously developed and cultivated lands are being abandoned. This means that the invaluable wealth, created by the work of preceding generations, is irreversibly lost. These processes are determined not by unfavorable natural conditions, but by the unfavorable socio-economic conditions of the life in the country. Characteristically, the process of rural-level exodus is restricted not only to remote regions, but it is typical of the very heart of the State, such as the Tver', Vladimir, Ivanov, Yaroslavl', Vologda, Kirov, Nizhny-Novgorod, Novgorod, Pskov, and other administrative regions, which have had formed part of central Russia's territory from time immemorial. The 'exodus' is even more characteristic for the regions of Siberia and the Russian Far East.

Another problem is the existence of depressed territories. They were formed during the course of restructuring of the key economic relations and the destroying the system of territorial organization of the economy, at local levels in particular. In Siberia this was determined by the fact that development in the past was driven by political motives associated mostly with the extraction and use of resources for heavy industry, and exploration of important strategic raw materials sources. Towns and settlements with a production and social infrastructure, were created at the geographical places of production and were entirely dependent upon the town-forming enterprises.

The system of settling and the structure of residential centers as basic objects of local government took shape and predominated in conditions of planned economy, centralized redistribution of all resources, and a common pricing mechanism. They come into conflict with the dynamics of the restructuring processes of the economy. The decentralization of the majority of managerial functions led to the dispersal and stratification of decision-making levels right down to a single individual. It compelled to reorganize the system of connections between society, business and authorities. In conditions of disintegration of sectoral management and privatization, there was an enhancement of the significance of the individual characteristics of economic agents connected with local conditions. The production and sales relations begin to form from the bottom up, and encompassing an ever increasing number of market agents. Therefore, there emerged real demand for the development of territorial planning and administration in the country, adequate to the new relations in society. Subsequently, in the new system of social organization the significance of the local level has increased dramatically.

The limited possibilities of Federal power as regards an effective administration of the local processes and the inadequate understanding of the transformation of the territorial structure of production and settling at a local level determined the current policy toward local government. The tendency is mostly towards transforming the local authorities into an active economic agent. Its goal now is to earn money to provide the required standard of living of a territory and supporting the present state with the limited transfers from higher-level budgets. The economic activities of local authorities in solving social problems leads to formation of business of economic formations with low efficiency on the basis of budgetary funds.

To overcome the policy, conserving the situation, the creation of an adequate system of goals is needed. With the basic functions of local authorities (law and order, public health, education) remaining unchanged, it contemplates the shifting assessment criteria for local authorities into the sphere of parameters, that reflect the involvement of the various local resources and the population itself into the creation of attractive living conditions and mode of life. Formally, this affects the development indices of the infrastructure that provides for material requirements of all branches of the economy of different forms of organization and property as well as informational needs, concerning the interests of entrepreneurs and population.

Consequently, Russian society is faced with many problems of restructuring the governing system for local socio-economic formations in the new conditions. To resolve these problems requires invoking the geographic science.

In the first place, this includes an improvement of the administrative and territorial managerial structure. The system of local self-government requires an optimization of spatial relations and boundaries of administrative formations in the interests both of management structures and of management objects.

Secondly, it is local markets investigation, which is needed as an informational basis for the development of enterprise as well as for the promotion of new productions.

Thirdly, it implies working out new functions of local self-government agencies related to the attitude toward natural environment. Under the planned economy, such functions were centralized and had a formal character. In the first instance, this applies for the principles of land use implementation at a local level as well as the exploitation of other resources, including ecological approaches in the regulation of the population activity. The State represented by local government agencies acts as a managerial institute and uses, in this case, legal, economic, organizational and administrative methods for environment control.

THE FORMATION OF A SANITARY-PROTECTIVE INFRASTRUCTURE

We are working out the avenues for the formation and development of a sanitary-protective infrastructure of residential centres. The infrastructure includes a whole package of measures for collection waste, rendering it harmless, and its recovery and for the control over industrial solid waste-flows.

The procurement-productive activity of the RF GOSSNAB (State Procurement Committee) was virtually liquidated in the first half of the 1990s. It was anticipated that the niche for procurement and processing of recoverable material resources would be filled by the business sector. Such was the case with metal resources due to the need for such a kind of waste in the market, in China in particular. The situation became complicated in view of the general unfavorable socio-economic situation in the country and region. Semi-legal and shadow businesses connected with the turnover of recoverable resources, especially nonferrous metals, emerged within a short period, and caused a great deal of damage to all branches of the economy and to the housing public utility sector. Authorities at all levels seem to be largely at fault in this case.

Formation of a sanitary-protective infrastructure for waste handling on a territory is closely connected with the involvement and use of its natural-economic potential and is to a certain extent an indicator of its socio-economic status and condition. The procurement block of the infrastructure in the south of Eastern Siberia encompassed almost all low-level administrative districts in the pre-reform period. During the period of market reforms, its procurement sector does function only in regional centers. Prospects for its rebirth are realistic at the level of administrative districts in the market-oriented conditions, but subject to significant investments and protectionist policy.

IMPROVEMENT OF THE SPATIAL INDICATION OF THE LOCAL-LEVEL INFORMATION BASE

Problems of rebuilding the relations in the country influence the development of the economic geography itself. The society information and improvement of technical means of management pose the task for science of creating the appropriate mechanism processes for collection and analysis information about the structure and functioning of territorial systems and decision-making on its basis. Geoinformation systems (GIS) aimed to provide services for certain levels of use and managerial functions could serve as the basic instrument among them. An improvement of the information base of a local level is one of the problems of GIS development. For solving the problem, it is important to start from a topological unification of official data, which is characterized by a gap in the territorial identification.

The unified national system for information support of management in Russia is built on maintaining a set of thematic, mutually reconciled and mutually complementing information bases. The State Land Cadastre that contains information about land uses, i.e. the location, purpose and legal status of the lands in the Russian Federation, about territorial areas and objects located on plots of land and closely connected with them would serve as a territorial basis for this system. Exactness of territorial localization of the land register data is limited by the adopted minimum size of the plots of land, while the qualitative component of the cadastre data is limited by the kind of property objects, mainly by sources for property taxation. Many of organs of State administration coordinate their activity in the information realm on the land cadastre base.

Besides the cadastre, the classifier of administrative-territorial formations also forms the basis for territorial identification; its data are used in many kinds of statistical information pertaining to information address referencing of juridical persons. To date the classifier has been developed for all administrative-territorial formations, except for residential centres. Thus the accuracy of the classifier territorial localization corresponds to the size of administrative units.

The land cadastre and the classifier are the only instruments for geographical referencing of official information in the State statistics system. The information about address referencing of juridical persons, contained in many statistical forms, cannot serve as a mean of informational and technological interaction of different information bases, because it cannot be arranged in the form of index fields.

Thus, referencing the existing information system to the territory is limited both territorially and substantively. These qualitative and territorial limits show up differently for different levels of State administration. When decisions are taken at the federal level, the limits have almost no influence on the quality of administrative actions, because the objects of management of this level are almost all presented either in the cadastre or in the classifier. At the regional level these limits start to have effect in the case where there is a need for exact information about a single economically or socially significant object, or there arises a problem of integrating information about it from different statistic data ases. At the local and municipal level the limitations of territorial identification are of the great significance because objects of management are most frequently point or areal objects, which do not fit into boundaries of administrative units or land cadastre units.

The mechanism for unifying statistical socio-economic information bases with regards to the location of objects of socio-economic activity can be improved in several ways. The easiest way is by driving the classifier to the level of a residential centre. This will then make possible to identify territorially the major part of official socio-economic statistical information, since that almost all economic activity of the population, enterprises and organizations is connected with residential centres. In this case the code assigned to a residential centre should meet the following condition: it should include the attribute that remains unchanged when the name, administrative subordination and the status of the residential centre is changed, and when settlements are unified and separated, that is, it must remain constant.

Only one of the settlement attributes that meet such requirements is the location, which is characterized by geographic coordinates. The mechanisms by which the geographic coordinates are fitted into the classifier's structure can be realized by adding several categories that reflect the coordinates of the residential centre, to the existing ten-grade index that determines the administrative rank of a settlement and its subordination. These "coordinate" index grades will be stable during all administrative transformations. This will permit keeping track of the uninterrupted dynamics of socio-economic information over time. Presently such a continuity is limited by any change in the administrative structure, leading to a change of values of the classifier's indexes.

The second avenue for integration according to the location of statistical data bases can be implemented by developing a special State classifier of economically or socially active transactors. Geographic coordinates should also form the basis for developing such a classifier. Initially, the State Plots of Land Register that contains data on coordinates of plots, on the one hand, and data on land use entities in the form of a classifier code of enterprises and organizations, on the other, would serve the role of such a classifier. Use of such data makes it possible to reconcile the land cadastre with other State statistical data bases, but only for those which are pertinent to land users. This falls for short of being the complete list of State economic subjects.

The development of the classifier according to the location ought to be started by formalizing the spatial forms of activity of juridical persons. The whole diversity of spatial referencing can be reduced to three main forms: point, area and linear. The area (areal) referencing can be continuous and fragmentary, structural, and hierarchical. The linear form can also be continuous and fragmentary with or without hierarchical structure. The point form can be unitary (one point) and multiple (several points). Thus the code (index) of subject location should contain the information about the spatial structure and hierarchy as well as geographical coordinates. This complicated geographical problem requires an extensive knowledge of the specific character and structure of the spaces that include quite a broad range of economical entities.

By developing such a location classifier, it could be possible to combine into a single whole, according to the location, the system of thematic information bases of State statistics.

INSTRUMENTS OF MANAGEMENT AND DIRECTED INFLUENCE ON THE DEVELOPMENT OF TERRITORIES

The interests of a local community should be reconciled with private and group interests. It is the lack of the mechanism of such reconciliation that hinders the evolutionary process of Russian society at the present time. A disruption in the system of territorial planning from bottom to top occurs at a regional level, which is accounted for by the status of low-level administrative districts in the system of economic relations of society. Low-level territories offering conditions and resources for economic activity, have no opportunities for their extended reproduction, in other words they do not earn money for it.

The key problem for development is the problem of creating instruments by which territorial managerial bodies stimulate the production activity. Out of the whole spectrum of control levers, we will single out the following three groups: 1) property, 2) territorial tax, and 3) agreement.

Most of the hope centered around the property of territorial formations (municipal and regional). From our point of view, the course of economic development and the tax reform has shown that this factor has not played an important role. It is able to give the income interest only but this is not enough for the formation of a significant investment potential, which can be used directly to solve problems of socio-economic development.

By territorial taxes is meant a set of rental payments and pollution fees, road and transport taxes, the sales tax, and a gamut of local taxes. It should be said that they are mostly limiting but not stimulant taxes. The promotion of development proceeds through a reduction of the territorial tax as regards own liability but it is normally of a minor nature and only rarely outstrips negative factors.

And the third regulating factor, the least tangible but, in our view, the most effective at the present time are contractual relations of local authorities with business partners. Territorial administration is effective only when it acts in a corporate form. Private capital would take actions desirable for territorial administration if they do not contradict its corporate plans and the results of such activity are of any material or moral benefit.

The mechanism for achieving the goals of territorial policy and solving problems facing individual territories may be arbitrarily divided into four main directions.

A traditional direction of the socio-economic policy focuses on the equalization of life conditions and aid to less developed territories. Such an approach is required for acutely depressive areas such as old-industrial, rural and northern regions. A calculation of the consumer basket and its value for different territories, according to which the policy of subsidies and other address measures is pursued, is one of the criteria of this difference and the instrument of regulation. An integrated approach to problems facing such territories can be implemented through target programs intended to resolve particular problems. Their funding should be included into the budget.

The second direction is based on agreement between authorities and sectoral industrial structures. General principles by which regional authorities influence the behavior of economic sectors in particular regions should be mentioned in this case. Measures of influence can be implemented not so much through the State or regional proportion of shares as through the conditions of productive resources: land with its natural potential, and manpower resources. At the present time such influence is of occasional or urgent character. The target-oriented work with an owner implies an a priori knowledge of the potential and purposes of development of a given territory. The oil and gas, aluminum, chemical and fuel industries are the objects of indirect influence on the territory of the Irkutsk region. The region can also have a direct influence on the development of lumbering and wood-working industries, the industry of building materials, and of the agro-industrial complex.

The integration policy implies a direct and target-oriented promotion of the exchange of commodities among local producers. This could be achieved by proceeding along two directions. In the first place, it is through the formation of intraregional markets, primarily food markets, which goes parallel with the sectoral policy but does not repeat it. It is built not so much on integration of agricultural enterprises with processing centers as on a further development of the market infrastructure for their own producers, and on the prevention of monopolization of both demand and supply. The second direction is the concentration of the economy around large industries, such as aluminum, oil and gas and wood-working industries. It is largely the territorial policy, as the region's old-industry territories are in a state of depression because of the shutdown of small processing and repairs-mechanical enterprises, and they have need for aid in establishing the relations.

The institutional direction implies administrative reorganization according to territorial problems, and an improvement of interbudgetary relations. The decrease in industrial potential of the old industrial towns and a relative stability of their food industry make their administrative integration with adjacent agricultural areas urgently needed to facilitate production and sales relations, especially in the conditions of the keen competition for stockbreeding products. Economy of the government expenses can also be achieved by concentrating certain administrative functions in several regions, centers of adjacent territories, even to the point of setting up 'okrugs'. The interbudgetary policy generally includes a traditional approach as regards the redistribution of income and the equalization of the conditions of life, yet another goal is equally important, namely, to improve the efficiency of budget expenditures and further the development of donor territories.

PART IV

METHODOLOGICAL ASPECTS OF GEOGRAPHY

CHAPTER

16

Landscape Planning in Siberia

A.N. ANTIPOV AND YU.M. SEMENOV

The most important conditions for an appropriate choice of a strategy for man-environment interactions with the environment to be correct include an error-free assessment of the natural-resource potential, including a complex estimation of geosystems and their transformation trends under the impact of both natural and anthropogenic factors. One of possible options to solve the problem of optimizing land-use and territorial planning for a stable regional development can be found in landscape planning (LP), widely employed in Germany and a number of other EU states.

Primarily, LP is a set of methodological tools applicable for building up a spatial organisation of the society's activities in specific landscapes, providing for a stable nature management and for the conservation of major functions of these landscapes as a life-support system; secondly, it is a communicative process enabling nature-managers' interests and nature-management problems to be revealed, conflicts to be solved and a consensual plan to be worked out for actions and measures to be taken, involving all subjects of nature-conservation and economic activities in the territories deemed for planning [1].

Despite the rather recent appearance of the LP concept in the Russian geographic literature, let alone in Russian nature-management practice, its traditions date fairly far back in our country. Thus, as early as the end of the 19th c., the Special Expedition of Russia's Forestry Department was organized, whose basic objective, in its chief V.V. Dokuchaev's opinion, was to work out integral, strictly systematic and consistent measures aimed at improving natural conditions for agriculture, as well as reguralarizing aquaculture. Practical measures were preceded by a detailed account of natural conditions by way of preliminary examinations performed in experimental areas: geodetic surveys, geologic, soil and climatic examinations. Widely known are also the researches carried out by Resettlement Bureau expeditions into Siberia and the Far East, including the fundamental mapping to supplement the component-wise assessment of natural conditions in the territories assigned for settlement by peasants from European Russia, as well as investigations performed by special expeditions whose purpose was to justify the deployment of productive forces in the USSR.

Unfortunately, the possibilities of using landscape-based land estimation and land-use planning are restricted by a lack of overall landscape surveys and relevant services. Moreover, such work, as well as the official procedure of ecological inspection for justifying land allotment and creating various production sites in Russia, does not appear to be a regular-type, continuous process of ecological maintenance for the project – from its conception to realization and subsequent control, – nor does it foresee a legislative finalisation of estimation/inspection results, and is virtually never applied but for some specific projects of economic activities [1].

Doubtless, the LP employed in Germany enjoys certain advantages in this respect compared to the above approaches. LP is implemented at varied scales and is consistent, spatial-level-wise, with the general territorial planning. The most important principle is the vertical continuity of target settings, with parallel concretization of tasks, measures and requirements, when descending from a higher planning level (landscape programs) to a lower one (landscape plans and lower).

Within the confines of the former USSR, projects similar to landscape planning were chiefly implemented in the Baltic republics – Lithuania and Estonia. For the Russian Federation, practical experience in similar work is virtually non-existent. Meanwhile, as was noted above, it was in Russia that the ideology and a number of methods for applied landscape research were worked out sufficiently completely, and a number of publications gave a characterization of the principles of geoecologic projection and socio-functional landscape analyses that are being developed by geographers-landscape researchers. In the 1980s, a form of regional ecological analysis was also developed known as "Territorial Complex Schemes for Nature Protection". Organizationally and compositionally, however, landscape planning best of all matches the widely known and widespread in Russia regional planning, whose principles and methods rely, to a certain extent, on landscape-ecological concepts, although it is exactly these principles that factual projects appear to have implemented rather formally and with insufficient effect.

In 1992, an Agreement was signed between the Russian Federation and the Federative Republic of Germany on cooperation in the field of environment protection. Within its frames, the Administration of Irkutsk Oblast* and the German Society for Technical Cooperation (GTZ) arranged to conduct a joint

Russian-German project in 1994, and landscape-planning activities headlined "Ecologically Oriented Land-Use Planning in the Prebaikalia"* were performed in the territory of Irkutsk oblast (Irkutsk and Olkhon districts) over the period of 1994-1998, within the framework of the Russian-German cooperation in landscape planning, by the Institute of Geography, Siberian Branch Russian Academy of Sciences (Irkutsk), assisted by the Institute of Geography, Russian Academy of Sciences (Moscow), Irkutsk Oblast Administration, German Society for Technical Cooperation (Eschborn, Germany) and the "Ecology+Environment" planning group (Hannover, Germany).

Russia's natural and social-economic conditions differing in principle from those in Germany, it was impossible to automatically transfer the German landscape-planning experience onto the Russian territory, requiring instead rather serious efforts to perfect planning methodologies. In Germany, where genuinely natural complexes are effectively absent, whereas the level of social relations in the nature-management sphere is particularly high, the planning's main objective consists in conserving natural bases of human activities, even at the expense of some land-user groups' interests. In the Lake Baikal zone (Prebaikalia), with its currently conspicuous degradation trends in the social-economic structure, the goal of planning appears to be more complicated, consisting instead in the necessity to respect the interests of local residents, while at the same time creating guarantees for conserving their natural environment.

Certain corrections were made to the methodological approaches proper. In Germany, territorial and landscape planning are two self-dependent processes that are, at a certain stage, integrated in one form or another. In Russia, current social-economic problems are so serious that unless they are taken into account, the realization of nature-protection measures starts to appear rather problematic. These realia are taken account of by way of a broader inclusion into the planning of social-economic factors of territorial development, wherefore, instead of making chiefly nature protection-oriented landscape plans, we, in our circumstances, had to produce ecologically-oriented plans for land use that focus largely on social-economic aspects of nature management.

Until now, such issues as the typology and legal status of territorial development plans have neither theoretically, nor practically, been properly worked out in the Russian Federation, which is also true of the planning procedure proper. In the Russian legislature, this sphere of legal regulation is characterized by many gaps and a lack of complex approach. The New Urban Development Code of the Russian Federation [2], though still not having exhaustively solved the problem, has to a certain extent regularised the system of urban-development plans, and specified the planning document titles. There now appear certain analogies to speak of in the hierarchy of planning documents defined by the Urban-Development Code of the RF. Meanwhile, given the spatial dimensions of planning objects, certain corrections should be applied to the scales of territorial analyses and generalizations.

Elaborating the planning methodology, the authors relied on the provision that LP is a tool for organizing the society's ecologically-expedient activities, its chief aim being not only to provide guarantees for the natural potential's long-term efficiency, but also to create guarantees for the local inhabitants' right to a decent life [3, 4].

LP is implemented as a hierarchical system where estimations, planning provisions and prescriptions of all levels are not mutually contradicting, but complementary instead, combining with each other according to the "countercurrents" principle, where frame recommendations (suggestions "from higher-up") serve as guidelines for more detailed directions at lower planning levels, while being themselves shaped under the effect of suggestions "from lower-down" (Table).

The LP procedure has a regular character and, normally, the making of landscape plans involves 5 major stages (Fig. 1):

a) inventory: collecting and generalizing all available data on the territory's natural environment, social-economic conditions, structure and land-use patterns, as well as uncovering major nature-management conflicts by analyzing the territory's imminent ecological problems;
b) estimating the planning territory's natural conditions and potential in terms of their importance and sensitivity; assessing land-use patterns;
c) elaborating component-wise target concepts of natural-resource usage for specific natural components;
d) elaborating an integrated target concept of territory usage;
e) working out a program for major directions of activities and measures to be taken.

To implement the project, 2 model land areas were selected: the River Goloustnaya basin with adjacent Lake Baikal shore; and the (Island of) Olkhon administrative district ("rayon") (Fig. 2). The R. Goloustnaya basin is a lowest-level territory of local (village) government. The Olkhon district is a higher-level territory of local (district) government where natural conditions, land-use patterns and the complex of ecological and social-economic problems are more diverse.

Plans for ecologically-oriented land use in the R. Goloustnaya basin [5] and Olkhon district [6] were frame plans. Their main purpose was to pinpoint principal landmarks for a stable social-economic development in the planning territories, given the sufficiently complete and tight ecological restraints proscribed by the Act on Lake Baikal Protection [7].

While developing and implementing the project resolutions of the frame plan for ecologically-oriented land use in Olkhon district, a package of cartographic documents has been compiled for the functional zoning of agricultural lands within the Pribaikalsky National Park, a landscape plan was composed for the area of chiefly recreational reclamation of Lake Baikal's Maloe More (the Smaller Sea) shoreline, and practical measures started to be taken aimed at improving the social-economic and ecological situation, as well as certain representative activities to be performed.

Table

Levels and scales of landscape and urban planning

Administrative level	Levels of urban planning	Landscape planning	Landscape-planning scale
RF subject, Group of subjects	Consolidated scheme of urban planning, Territorial complex scheme	Landscape program	1:1,000,000 to 1:200,000
District, Group of districts	Territorial complex scheme of urban planning	Landscape frame plan	1:200,000 to 1:50,000
Local government territory, Large city, Specially protected territory	Territorial complex scheme of urban planning	Landscape plan	1:50,000 to 1:25,000
Settlement, City community, SPT section, Part of a territory under local government.	General plans	"Green" plan	1:25,000 to 1:5,000

Inventory

Collecting available information about the state of natural environment and socio-economic conditions of the territory	Identifying major conflicts and problems in the sphere of nature management and socio-economic relations	Determining the natural environment components for analysis, and priority natural component functions of development

Assessment

Developing the criteria for the assessment scale of sensitivity and significance of separate natural components	Territory zoning according to the significance and sensitivity of separate natural components for implementation of the target component function

Working out the component targets of territorial development

Developing the concept of integrating assessment categories into territorial component targets of development	Territory zoning according to component targets

Integrated concept of targets of territorial development

Developing the concept of integrating the component targets of territorial development	Territory zoning according to integrated targets

Concept of major directions of action and priority measures

Integrating land use conflicts with the concept of targets	Defining the types of measures and zoning them	Defining the measures according to economic sectors

Figure 1: Landscape frame plan development stages (in more detail).

In subsequent years, the SB RAS Institute of Geography, with active advice support from the German Federal Agency for Nature Protection, went on to develop a methodological apparatus, chiefly by way of widely employing landscape-planning methods along various directions of nature protection in the territory of the Prebaikalia. A 1:1,000,000-scale scheme of ecological zoning of the Lake Baikal Natural Territory was developed for the first time in Russia [8], to become a first document in the implementation of the Russian Federation Act on Lake Baikal Protection. Also developed were landscape frame plans for Slyudyanka district [9] and southern Irkutsk district, the R.Selenga basin [10] and the Zabaikalsky National Park, as well as large-scale landscape plans for the village of Listvyanka and the city of Baikalsk [11] (see Fig. 2). This research became a basis for developing principles of landscape planning and its development concept in Russia and for elaborating methodological recommendations for landscape planning [1, 12].

The specific feature of a landscape program as a higher hierarchical level in the landscape-planning system consists in marking out major functional (target) usage zones in the whole of the planning territory, while conserving Lake Baikal's unique ecological system and preempting any negative impacts on its health from economic and other activities. Its development is based on analyses of the relevant nature's spatial structure, anthropogenic load on geosystems, current and potential restraints on economic activities [8]. The main objective of the landscape program for the Lake Baikal natural territory consists in setting off types of ecological territories regulating in some way the economic activities. This could only be achieved by simultaneously solving two problems – conserving the Lake Baikal natural territory as a World Natural Heritage site and providing for this territory's stable social-economic development without infringing the rights and liberties of its inhabitants.

Therefore, the program relies on a territorially differentiated approach to setting out types of ecological territories (zones). They constitute geographic systems of varying environment-forming importance and resistance to anthropogenic stress. They are differently affected by objects of economic activities modifying and transforming the natural geographic systems or specific landscape components.

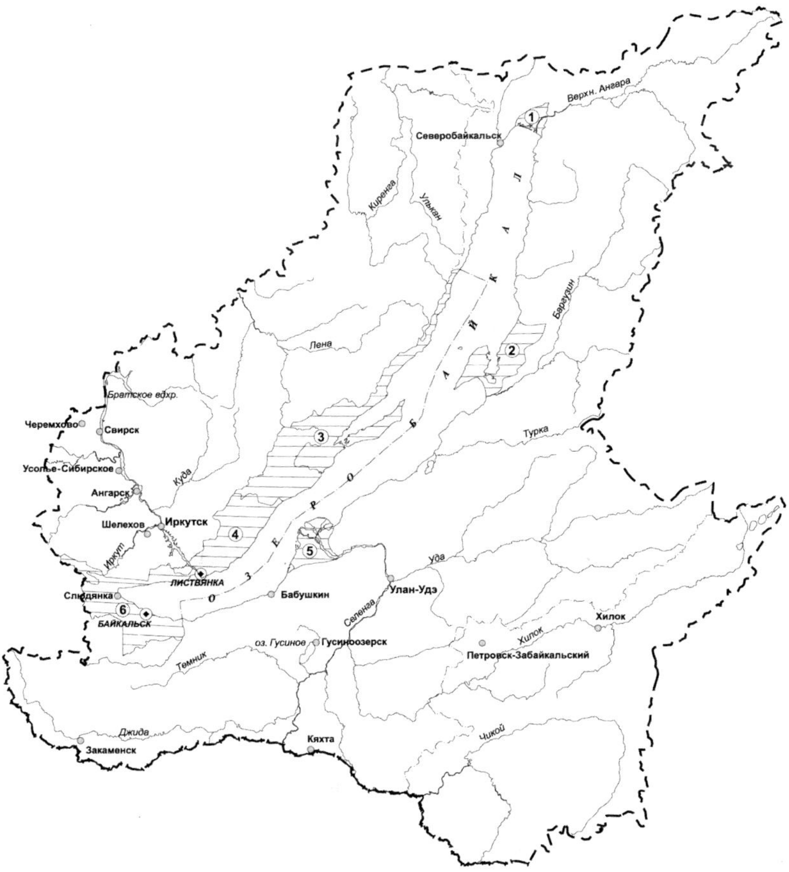

Landscape programme for Lake Baikal Natural Territory development

Frame landscape plans:
1 – R.Verkhnyaya Angara delta
2 – Zabaikalsky national park
3 – Olkhon district
4 – R.Goloustnaya basin and south-western Irkustk district
5 – R.Selenga delta
6 – Slyudyanka district

Large-scale landscape plans

Figure 2: A scheme of model areas within the Lake Baikal Natural Territory.

Three ecological zones have been recognized in the LBN territory (Fig. 3): the Central Zone, including Lake Baikal with islands, adjoining Lake Baikal water-protection zone and specially protected natural territories adjoining Lake Baikal; the Buffer Zone – territory beyond the central ecological zone, including Lake Baikal's catchment area within the boundaries of the Russian Federation; the Zone of Atmospheric Influence – the territory beyond Lake Baikal's catchment area within the boundaries of the Russian Federation, expanding up to 200 km westwards and north-westwards.

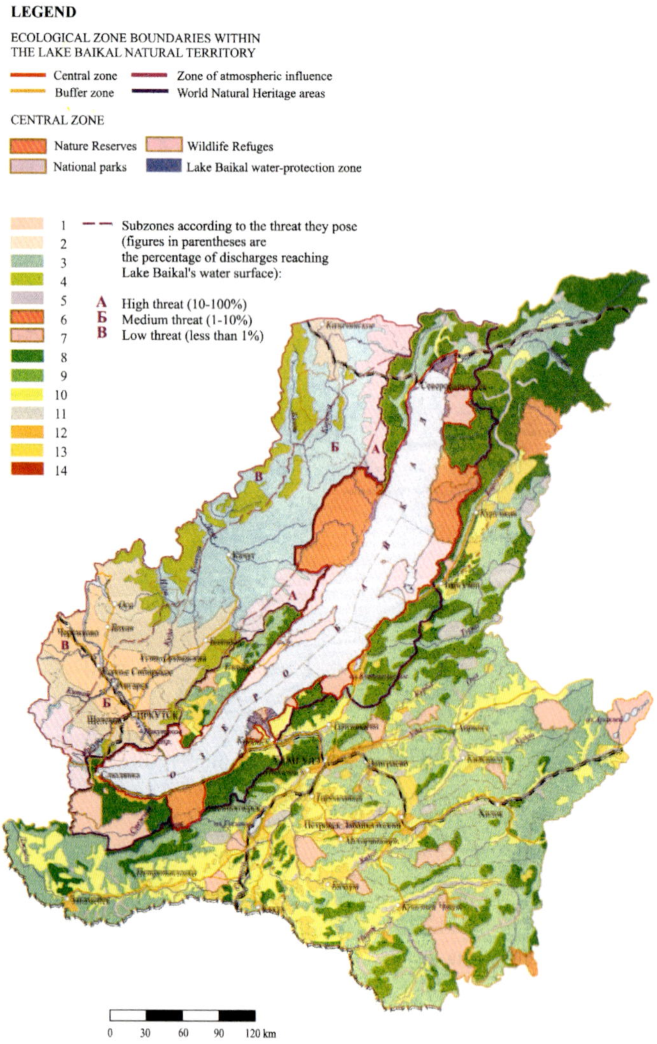

Figure 3: Ecological zoning of the Lake Baikal Natural Territory.

With large-scale LP at a communal level, its goals are generally not oriented towards solving general problems of territorial development, addressing instead specific priority problems defined, in the context of nature-protection demands, in frame prescriptions of higher-level plans.

Zone of atmospheric influence

No	Climate type	Protection and rational usage of climate resources for territory development
1	R.Angara plain-valley	Atmospheric situation improvement, to progress to the Regulated Intense Development category
2	Foothill plain-valley	Regulated intense development
3	Valley-trough and narrow valley	Extensive recreational and agricultural development
4	Plateau	Conservation while proceeding with the existing extensive usage
5	Medium- and high-altitude mountains	Conservation of current state, any usage denied

Buffer zone

No	SUBZONES	TERRITORIES	LANDSCAPES	CURRENT USAGE	SUGGESTE DUSAGE
6	Nature Reserve regime	Nature Reserves	Highlands: bald peaks and taiga-hill	Usage denied	Usage denied
7	Wildlife Refuge regime	Wildlife Refuges, national parks	Bald peaks, taiga, forest-steppe, steppe, lake-marsh	Usage temporarily denied	Recreation
8	Top-priority conservation; undeveloped	Highlands; territories remote from inhabited areas	Bald peaks; steep-slope taiga with highly active steep-slope processes	Hunting, tourism	Recreation, hunting, Ecological tourism
9	Existing management (mainly forestry)	Forest	Medium- and high-altitude mountains with low-gradient hill-taiga; foothill	Primary and secondary forest usage	Forest management with areas under natural, historical and cultural monument protection regimes

10	Developing management (agriculture, transport)	Forest-steppe, steppe, territories close to populated areas	"Steppe-ified" trough bottoms; below-taiga	Agriculture, cattle-breeding, transport	Developing current sectors; gas-pipe and power-line construction
11	Regulated intense development	Major developed lands, mineral deposit sites	Below-taiga; foothill steppe; valley	Industrial production, mining, transport, agriculture	Industrial, mining, transport, agricultural, therapeutical and sanative
12	Improving ecological situation in industrial centres	Industrial centres, erosion-endangered agricultural lands	Transformed valley	Industrial, mining, transport	Industrial, mining, transport
13	Forest restoration, land recultivation and enhancement	Forest clearings and burns, mining waste sites, ploughed lands	Disturbed taiga, transformed valley	Forestry and agriculture	Forestry, recreation, agriculture
14	Landscape restoration, to progress to the Wildlife Refuge regime category	Wind-blown sands; natural, historical and cultural monuments	Valley, wind-blown sands	Agriculture	Recreation; usage temporarily denied

A major aspect of large-scale LP is the conjugate analysis of problems in the spheres of politics, social welfare, economy and ecology – as applied to the supreme goal of developing a particular territory, defined in the process of frame planning and subsequent analyses of the goal tree. Each of the above spheres is subdivided into specialized elements. Thus, the sphere of e.g. economics is further broken down into the processing industry, forestry and aquaculture, transport, power generation, and commerce.

To take into account particular natural-ecological conditions covered by a large-scale LP, to evaluate and elaborate the target development settings, one selects the most significant nature components. For the given scale, they most frequently appear to be soils, species and biotopes, landscapes and their recreational potential.

The Institute of Geography has gained experience in using LP for solving particular tasks: land-use planning, water-protection zoning, urban-development planning and estimations of the environmental impact of economic objects to be created. It is at a large-scale level that LP is practicable in solving specific tasks.

As was noted above, functional zoning of agricultural lands was performed for the territory of Olkhon district, Irkutsk oblast. LP methods applied to land-use planning provide facilities for correctly estimating the land potential and for the land's functional zoning, distinguishing different-purpose lands in terms of estimating their quality, sensitivity to anthropogenic loads, current usage, and defining target functions for further development. The scheme of the lands' functional zoning was made to a scale of 1:25,000, the basic scale for land-use planning documents.

The functional zoning procedure heavily relies on the map of actual land-usage, drawn up to show the territorial and natural-economic structure of agrarian land-use, distribution of land stock between land users, its productive specialization, distribution of settlements and production sites (cattle-farms), the population's ethnic composition. The map is divided into segments devoted to particular cadastral zones – integral territories controlled by a particular self-government organ, including all categories of lands (those deemed for agriculture, settlement, industry, power generations, transport, communication, radio and TV broadcasting, informatics, cosmic support, defense, security, other special purposes, specially protected territories and objects, forest stock, water stock) and land users, irrespective of their ministerial or any other allegiance. The contents of each map of actual land-use, depicting an agroterritory subject to a particular village administration, are reflected in a separate individualized legend.

For the natural-ecological conditions to be included in territorial analyses and in the estimation and elaboration of development target settings done using LP methodology in order to justify land-planning, the most significant natural environments appear to be species and biotopes, soils, landscapes and recreational potential. The functional zoning scheme of agricultural lands in Olkhon district, Irkutsk oblast, within the bounds of the Pribaikalsky National Park, combines the zoning of types of territorial development targets and of the territory's economic functions.

The current urban development activities take into account ecological and natural features of settlement territories and aim at providing favourable conditions for predicting and controlling the nature management and environment protection. In the process, the problem of a stable development of both settlements and between-settlement territories is being solved, including restricting the harmful effects of economic and other activities on the natural environment and its use in the interests of both the present and future generations. It is the application of LP approaches that best of all promotes the implementation of thus raised general issues. Landscape-planning work performed for the city of Baikalsk provides an example [11].

The first stage in this planning involved a water-protection and complex functional ecological zoning of the territory of Baikalsk and its nearest environment using evaluation and analytical 1:10,000 and 1:25,000-scale maps, the city and its environment treated as an integral territory. Territorial development goals were further defined for the outlined zones using LP technologies and bearing in mind the current normative legal ecological restraints imposed by the federal and regional legislation.

The prepared series of cartographic material included 18 maps. The resultant material for analyzing problems of the city's development consisted of synthesized maps of territorial development goals for superficial waters and biota, as well as two integral maps of an even higher generalization level – integrated goals of territorial development and integral zoning as per kind of major measures to be taken. To realize the uncovered categories of integrated goals in territorial development, principal directions were suggested for actions and measures to be taken. For the entire territory, 19 areas were marked off characterized by their development goals, with a general description of their nature structure and human-induced modifications in it (geographic localization). For these areas, actions and concrete measures were outlined whose realization is indispensable when trying to achieve the goals of the territorial development for particular land areas.

Therefore, one may conclude that the Russian-German cooperation in Russia resulted in laying down corner-stones for a wide implementation of landscape-planning tools as a basis for a stable territorial development. Given the proper legal finalization of the procedure and results of such activities, the country's territorial problems could be solved while at the same time conserving its unique natural riches without damaging its socio-economic status.

References

1. Landscape Planning: Principles, Methods, European and Russian Experience. Bonn-Moscow – Irkutsk, 2002.
2. Urban Development Code of the Russian Federation / Rossiyskaya gazeta, 1998, No. 73.
3. Antipov A.N., Kravchenko V.V., Semenov Yu.M. Landscape planning in the Prebaikalia. Geografiya i prirod. Resursy, 1997, No. 4.
4. Antipov A., Hoppenstedt A., Kravtschenko V., Semenov Ju. Oekologisch orientierte Landnutzungsplanung in der Baikal-Region. Garten und Landschaft. – Muenchen, 1997, Heft 9.
5. Ecologically Oriented Planning of Land Use in the Lake Baikal Region. The R. Goloustnaya Basin: a 1:200,000-Scale Frame Plan of Ecologically Oriented Land Use. Irkutsk – Hannover, 1997.
6. Ecologically Oriented Land-Use planning in the Lake Baikal Region. Olkhon District: a 1:200,000-Scale Frame Plan for Ecologically Oriented Land Use. Irkutsk, 1998.
7. Federal Act of the Russian Federation (1 May 1999 No. 94-FZ) "On the protection of Lake Baikal", Rossiyskaya gazeta, 1999, No. 36.

8. Ecologically Oriented Land Use Planning. Ecological Zoning of the Lake Baikal Natural Territory. – Irkutsk, 2002.
9. Ecologically Oriented Land Use Planning in the Lake Baikal Region. Slyudyanka District. Irkutsk, 2002.
10. Ecologically Oriented Land-Use Planning. The R.Selenga Basin. Irkutsk, 2002.
11. Ecologically Oriented Land-Use Planning. The territorial development of the city of Baikalsk and its natural zone. Irkutsk, 2003.
12. Guide-Book on Landscape Planning. In 2 vol.: Vol.I. Principles of Landscape Planning and the Concept of its Development in Russia; Vol.II. Methodological Recommendations for Landscape Planning. Moscow: State Centre for Ecological Programmes, 2000–2001.

CHAPTER

17

Thematic Mapping of Siberia

A.R. BATUYEV

Siberia's map-making dates back to its annexation to Russia. Based on written and graphical records from the Epoch of Great Geographical Discoveries, the earliest maps of Siberia were produced – they were called "drawings" at that time. Noteworthy among them are "The Drawing of Siberia" by P.I.Godunov, and "The Drawing of Siberia to the Chinese Kingdom…" (1667). These two maps are the earliest of all national generalizing maps of the 17th century that have persisted to date, and provide evidence of the extensive mapping efforts made by the Russian State under Aleksei Ivanovich' rule. In the case of "The Drawing of Siberia" (1667) the very first attempt (in national cartography as documented in scientific records) was made to unify map symbols [1].

A breakthrough in all national cartography before Peter the Great was represented by the remarkable works by a talented Siberian man of science S.U.Remezov (1962 – after 1720), who authored a series of publications on geography, ethnography and history of Siberia. S.U.Remezov compiled three collections of maps, sketches and drawings: "The Drawing Book on Siberia" (1699-1701), "The Chorographic Drawing Book" (1697-1711), and "The Service Drawing Book" (1702-1703), which were completed by his sons. S.U.Remezov was entrusted with the compilation of the first of these atlases, "The Drawing Book on Siberia", in accordance with the statutory act of the Russian Government, the *boyar* "judgment" of 10 January 1696.

General maps of Siberia and of its separate regions in "The Service Book" and "The Drawing Book", combined with varied primary topographic materials assembled in the "Chorographic Book", provided for the first time in history of national and international science a reliable and detailed historical-geographical and cartographic description of the Siberian territories. Those first atlases of Siberia are amazingly attractive for detail of terrain representation, and for the volume of historical and geographical information displayed. In His publications, S.U.Remezov formulated a special style of traditional Russian cartography, very high (for those times) "standards" on the volume of information represented by the map. At early stages of survey by geodesists at the epoch of Peter the Great, not only were "those standards" outstripped, but they were to a large extent minimized.

After the epoch of Peter the Great, the study of Siberia was placed on a scientific footing; this is primarily true for cartography that was firmly established for its further development (instrumental survey of the terrain on an exact mathematical basis). A rapid advancement of commerce and industry during the reforms initiated by Peter the Great dictated the need for a special setting of the cartographic study of the country. Maps were becoming engineering documents with accurately determined metric properties, and were supported by a constant scale, projection and geographical coordinates. Compilers of the old Russian drawings had no adequate professional geodetic or cartographic training.

The rich national traditions of Russian cartography that were laid down by S.U.Remezov as regards the completeness and reliability of the contents of maps were not forgotten and were further developed in the works by I.K.Kirilov, V.N.Tatishchev and M.V.Lomonosov. A breakthrough was represented by the publication of such fundamental works as "The Atlas of the All-Russia Empire" by I.K. Kirilov (1734), and "The Atlas of Russia" (1745) which, for the first time, Siberia and the Far East were portrayed in accordance with their true dimensions and outlines.

For a further development of cartography in the mainstream of Russian survey traditions, a positive role was played by extensive mapping documents of general land chorometry, and by the institution of the military-topographic service in Russia during 1765-1822 (the creation in 1797 of the Depot of Maps, and in 1822 of the Corps of Military Topographers). The need for information about the geographical placing of various projects within provinces and regions rules by governor-general was enhanced especially during the implementation of th administrative reform of 1775. Those data were all to be presented by land-surveyors of *guberniyas* and *uyezds* (administrative units), who in the mid-70th-80th of the 18th century carried out extensive work on establishing and specifying the boundaries of various administrative-territorial units as well as their detailed geographical description, the compilation of various economic-geographical statements, lists of human settlements, etc. In Siberia in this case there was a common practice of compiling (according to assignments from central or local administrations) various, frequently schematic maps, having an important operational-reference significance for local authorities. The Corps of Military Cartographers in Siberia, as well as throughout Russia, was in charge of organizing cartographic activities on accurate topographic bases. The Corps of Civil Cartographers was instituted in 1838 in the newly established Ministry of State Properties (MSP) "for land-surveying and inventory of State-owned lands and areas of economic significance". The MSP paid special attention to organizing work on the study and map-making of State-owned lands of Siberia.

The development of the All-Russia market in the 19th – beginning of the 20th centuries, the increasingly intensive involvement of the country in international commodity turnover, and specialization and concentration of production dictated the need for an extensive deployment of cartographic work in order to meet the requirements of transport, industry, and agriculture and to study and extract mineral resources in the country's vast regions. At that period, various agencies in Siberia used to compile thematic maps of quite different kinds, most of which have persisted as manuscripts to date. Overall, the results from studying and mapping Siberia and the Far East before the October Revolution of 1917 are presented in a fundamental three-volume monograph "Asian Russia" and portrayed in "The Atlas of Asian Russia" (1914). Together with extensive explanatory text, three volumes of this atlas provided, for the first time, in a summarized form a graphical characteristic of Siberia, the Far East, and Middle Asia.

After the October Revolution the accelerated development of the country's eastern regions dictated the need for an enhancement of scientific-research exploration, on a cartographic basis, by various State ministries and agencies, including the Academy of Sciences, and Institutions of higher education. They organized large-scale expeditions: to the Indigirska (1926), Taimyr (1928-1929), Yakutia (1925-1934), Altai (1927-1928), the Kolyma (1929-1930), Kulunda (1931-1934), Chukotka (1933-1934), Oirot (1936-1937), and to other places, which promoted the development of productive forces in many regions of Siberia and the Far East. Experience gained by doing that research set the stage for the institution and deployment (in many regions of Siberia and the Far East) of field surveying cartographic work at a large scale. Subsequently, the implementation of major regional economic programs, such as development of the oil and gas resources of West Siberia, the creation of industrial and energy complexes in the Angara-Yenisei region, including the creation of KATEK, comprehensive development of the natural resources along the BAM, etc. called for a relevant cartographic support with regards to the goals and tasks conditioned by the specific character of the programs themselves, their phases, and by regional adjustment.

The process of fundamental scientific geographic mapping in Siberia has progressed through several major stages. As its first stage one could name the complex inventory-interpretation mapping of nature, economy and population, i.e. complex mapping proper. The beginning of this stage was marked by the compilation of the Atlas of Irkutsk oblast (1962), a joint work by geographers and cartographers of Siberian M.V. Lomonosov Moscow State University.

To coordinate work on thematic cartography issues, in 1964 V.B. Sochava initiated the appointment of a commission for integrated mapping of the natural environment, economy and population. The Commission functioned on the Institute of Geography's premises, but was under the jurisdiction of the RAS Siberian Branch and performed functions of a Siberia-wide combine [2]. It was under the Commission's auspices that, periodically, scientific conferences on cartography were held and complex pieces of cartography published as fundamental regional and national atlases or desktop-map series (Table 1), as well as in the form of multi-sheet scientific reference wall maps (Table 2).

Major works of the Institute in the field of atlas-making (see Table 1) include the atlases "Transbaikalia", of Amur oblast, Lake Hovsgol, and the National National Atlas of Mongolia. The atlas "Transbaikalia" was published in 1967. An original program of the entire atlas, a number of new subjects and techniques employed in charting (particularly as regards vegetation, landscapes, agriculture, population, etc.) attached to the Atlas not only a regional but also general scientific significance. For the first time in national cartography, the atlas "Transbaikalia" provided a detailed description of Lake Baikal. The section on Baikal aroused such a great interest that in 1969 it was published as a separate atlas of this lake.

The largest of the atlas products is the National Atlas of Mongolia. In this atlas, staff members of the Institute of Geography SB RAS, in collaboration with Mongolian specialists, have fully provided for the creation of all socio-economic maps and relevant texts. Its 17 thematic chapters outline the country's territory, natural conditions and resources, patterns of its economy's sectoral, social and territorial structures, population problems, man-society-nature interactions, and environmental health. The Atlas has a first-priority importance for the inventory of natural and labor resources, the whole of Mongolia's productive potential.

Multi-sheet wall maps display the territories of the whole of Asian Russia, southern East Siberia, West-Siberian plains, the Baikal-Amur main railroad (BAM), the Republic of Buryatia, Irkutsk oblast, and others. The largest of the series is the map series for southern East Siberia (see Table 2). Produced to the same scale (1: 1,500,000), these maps, taken together, provide a large amount of fundamental scientific information about this important region. They are supplemented with a series of 9 agroclimatic maps (1968), a soil map of Irkutsk oblast (1988) and a series of ecological maps of the Republic of Buryatia (1993, 1996, 1998).

The second stage of elaborating complex mapping paid prior attention to cartographic support for developing the sectoral and territorial production complexes. In 1972–1977, the Institute produced a map series of agriculture in southern Krasnoyarsk Territory, a model of developments accomplished in the interests of a particular sector of the economy. Another example of cartographic special-purpose products is provided by a map series for the Kansk-Achinsk fuel-and-energy complex (KATEK). It belongs among Siberian geographers' products aimed at providing scientific support for plans and projects dealing with the economic development of Siberia's regions. The purpose of the work is to study the prospects for creating the KATEK, given the limiting factors of natural environment, and working out recommendations for relevant nature conservation. The series includes groups of maps characterizing the natural and social-economic conditions in the region, its resources, factors and sources of anthropogenic impacts on environment, and, finally, predictive and recommendation maps that provide an idea of the trends occurring in the geographic environment, and suggesting measures for its optimization.

Table 1

Regional and national atlases, desktop and electronic scientific-reference map series

No.	Product title	Participants	General scale	Year of publication
1	Atlas of Irkutsk oblast	Moscow University	1:4,000,000	1962
2	Transbaikalia. Atlas		1:3,500,000	1967
3	Southern Krasnoyark Territory. A series of maps for agriculture planning (39 maps)		1:750,000	1977
4	Atlas of Amur oblast		1:2,500,000	1981
5	Yakut ASSR	University of Yakutia	1:5,000,000	1981
6	Socialist Yakutia	University of Yakutia	1:5,000, 000	1982
7	Amur oblast. Hydroclimatic resources (15 maps)		1:5,000,000	1983
8	Atlas of Lake Hovsgol	Irkutsk and Mongolian Universities	1:1,000,000	1989
9	Kansk-Achinsk fuel-and-energy complex (edition 1, 46 maps)		1:2,500,000	1991
10	Cartographic image of Irkutsk oblast and Lake Baikal region (CD ROM, 44 maps)		1:2,500,000	2002
11	Atlas "Irkutsk Region: Ecological Conditions of Development" (175 maps)		1:2,500,000	2004
12	Atlas "Sustainable Development of the Baikal Region" (electronic atlas)		1:1,000,000	2006

Table 2

Multi-sheet scientific-reference wall maps and map series

No.	Product title	Participants	General scale	Year of publication
1	Geobotanical map of the Ob-Irtysh taiga in West Siberia		1:2,500,000	1972
2	Hypsometric map of the Ob North ("Obsky Sever")		1:2,500,000	1972
3	Bathymetric map of southern Kara Sea		1:1,500,000	1976
4	Vegetation of the West-Siberian plain		1:1,500,000	1976
5	Hypsometric map of the West-Siberian plain		1:1,500,000	1982
6	Baikal-Amur railroad Vegetation: Main functions vegetation Vegetation cover		1:2,500,000 1:2,500,000	1983 1983
7	Soil map for Irkutsk oblast	Siberian Institute of Plant Physiology and Biochemistry SB RAS	1:1,500,000	1988
8	Landscape-ecological map of Buryatia	Baikalsk Institute of Nature Management SB RAS	1:1,000, 000	1993
9	Ecological-economic map of Buryatia	Baikalsk Institute of Nature Management SB RAS	1:1,000,000	1996
10	Ecological-geographical map of the Russian Federation	Moscow and St.Petersburg Universities	1:4,000,000	1997
11	Nature management of Buryatia	Baikalsk Institute of Nature Management SB RAS	1:1,000,000	1998
12	Series of agroclimatic maps of Irkutsk and Chita oblasts, and Buryat ASSR (9 maps)		1:2,500,000	1968
13	Series of maps of natural environment, population and economy of Asian		1:8,000,000 1:7,500,000 1:7,500,000	1974 1977 1988

	Russia: Red ticks Correlation ecological-phytocenotic Rodents and Lagomorpha			
14	Series of maps of natural environment, population and economy of southern East Siberia Economic Vegetation Landscapes Population Land utilization		1:1,500,000 1:1,500,000 1:1,500,000 1:1,500,000 1:1,500,000	1968 1972 1977 1979 1988
15	Ecological zoning of the Baikal natural territory (two0sheet montage of the main zoning map, and 4 map insertions	Baikalsk Institute of Nature Management SB RAS Institute of Natural Resources, Ecology and Cryology SB RAS	1:1,000,000	2002

The progress of work in producing specific maps and atlases has prompted the Institute's geographers to address various issues of thematic cartography theory, and, simultaneously, methodology problems. Issues of geographical cartography theory and methodology were dealt with in more than 20 large monographs and thematic collections published by the Institute's staff [1]. Among them, one should first of all mention: "Vegetation Cover on Thematic Maps" by V.B. Sochava (1979); "Cartographic Methods of Exploring Geographic problems in Agriculture" (1970) by V.P. Shotsky; "The Field Concept in Modern Cartography" (1978) by V.A. Chervyakov; "The National Atlas of the People's Republic of Mongolia. Problems and Scientific Contents" (ed. by V.V. Vorobyev, 1989); "Producing the Lake Hovsgoll Atlas (experience in complex mapping)" (1996) by B.A. Bogoyavlensky, A.G. Zolotarev, O.M. Kozhova et al.; "Ecological Mapping of Siberia" (1996) by V.V. Vorobyev, A.R. Batuyev, A.V. Belov et al.; "The Regional Ecological Atlas: Concept, Problems, Scientific Contents" (1998) by A.R. Batuyev, A.V. Belov, V.V. Vorobyev et al.; "Geosystems and Mapping of the Ecological-Geographic Situation in the Near-Selenga Depressions in the Lake Baikal Region" (2000) by A.R. Batuyev, A.B. Buyantuev, V.A. Snytko; "Cartographic Studies of the Biota" (2002) by A.V. Belov, V.F. Lyamkin, L.P.

Sokolova; "Methodology of System Ecological Cartography" (ed. by V.V. Vorobyev and V.A. Snytko, 2002), and others. The Institute's scientific journals have published papers addressing theoretical and methodological issues of cartography, all of which had a parallel ultimate practicable objective.

A noticeable amount of complex cartographic work has been done by expeditions. Many maps and sketch-maps with economic-geographic contents were composed in East Siberia (Irkutsk and Chita oblasts, Krasnoyarsk Territory and the Republic of Buryatia). These maps were a response to demands that emerged in connection with the creation of the Bratsk and Ust-Ilimsk industrial complexes, the Irkutsk – Cheremkhovo industrial and the Sayansk energy centers, and the construction of the BAM and KATEK. Maps were also compiled in connection with small-scale regionalization, and other socio-economic tasks.

West Siberia has also been an object of multi-purpose geographic mapping regarding problems of its development. Special attention was paid to the plains' north – "Obsky Sever" (i.e. the r Ob North), – for which a whole set of nature, nature-economic (resource) and economic-geographic maps has been created.

In the Transbaikalia's mountains, major prospects for human life and activities are concentrated in troughs. Therefore, the Institute focused its attention on comprehensive studies of the Upper-Chara and Barguzin depressions. For each of the them, medium- and large-scale thematic map series have been compiled. Similar maps were partly made for the Muya and Upper-Angara depressions. Soil examinations were carried out there, and soil maps drawn, permiting, together with the other indices, judgments about the prospects of plantations. The Institute's scientists have produced maps of geosystems and natural resources for the near-Selenga depressions in the Transbaikalia.

Over the past few years, a markedly intense development was observed in geographic cartography in such fields as mapping of recreational activities, cartographic support of landscape planning and ecological zoning in the Lake Baikal region, and ecological support of large economic projects in various regions of Siberia and adjacent territories in the Far East and Central Asia. These new cartographic products are of fundamental general scientific significance with important practical implications.

Overall, the second stage occurred when the development of ideas and methodologies of a complex approach prompted synthetic cartography to be formed, bringing forward the integral imaging of regional systems and structures by integrating their components' parameters, properties and interrelations. A specific role was played by the assessment-forecast cartography, whose aim is to purposefully and integrally assess the development of countries and regions. The end of this stage saw the next step – the development of the system cartography, focusing on integral imaging of geosystems and their elements, hierarchy, interrelations, dynamics, and functioning. The system approach worked well, on the one hand, in the new methods of cartographic modeling of geosystems, and, on the other, in systemic organization of the cartographic process proper.

An obvious fact at this stage is the presence of another, equally important, direction in comprehensive geographic cartography: cartographic support of regional development programs, i.e. problem-oriented cartography. During 1970–1980, the range of problems in ecological cartography attained an especially vital character.

Much attention is paid by Siberia's cartographers to the ecological situation in the largest cities of Siberia. They produces a sequence of ecological maps for the city of Irkutsk and its suburban zone (V.B.Sochava Institute of Geography SB RAS), the electronic ecological atlas of Novosibirsk (GIS Center SB RAS), ecological maps of Krasnoyarsk (Krasnoyarsk Scientific Center SB RAS) and Kemerovo (Kemerovo Scientific Center, and the Institute of Coal SB RAS), and others.

The V.B.Sochava Institute of Geography SB RAS, together with Moscow and St.Petersburg Universities, compiled the ecological-geographical general map of Russia to a scale of 1:4,000,000 [3], and, together with the Baikalsk Institute of Nature Management SB RAS, the map of ecological zoning of the Baikal natural territory to a scale of 1:1,000,000.

The Institute's major cartographic work along this line of scientific inquiry is the atlas "The Irkutsk Region: Ecological Conditions of Development" that was published in 2004. It was conceived, developed and produced not as an ordinary regional product but as a fundamental cartographic document, reflecting the interrelations between social production and natural environment in all its complexity, as well as the interactions between internal and external factors. The purpose of this fundamental atlas is to provide an opportunity to reconcile the ecological links and phenomena portrayed on the maps with a broad geographical background of a varied thematic spectrum, as well as to display the particular indices of some or other deviations in the nature management system in order to eliminate them by carrying out measures on its rationalization. Different-level authorities and agencies, scientific, educational, designing, production and other institutions, and public organizations and movements will be provided with reliable material to be used in generating scientific, technical and other projects, programs and initiatives of territorial development with due regard for the resource and ecological component. The atlas contains 175 maps which are supplemented by diagrams, plots, schematics, tables, photographic maps, and color photographs. Explanatory text in the form of scientific articles amounts to 10 printed sheets. Concurrent with the publication of a classical (printed) version of the atlas, its electronic CD ROM version is available, and a digital version of the atlas has been developed in the form of a regional GIS. The digital version of the atlas, with all capabilities taken into account, represents not only a set of illustrative cartographic documents, but is becomes an aggregate of interrelated data bases generated on a common conceptual and methodological basis.

The synthetic and systemic directions in cartography, and also the immediate tasks of solving the vital problems of regional development by means of cartography required a substantive development of mathematical methods and automated computer-aided technologies of map-making. In this way, a step was made towards creating automated cartographic systems and GIS's.

The advent and further development of new methods of map-making as a direct extension of system and problem cartography in a new information environment gave rise to a third stage, geoinformation cartography (GC), organized as a corel discipline at the intersection of automated cartography and GIS's, system cartography and aerospace methods in a broad sense of the word, including remote sensing, interpretation and digital photogrammetry. A major impetus for the emergence and development of GC as a core discipline was given by the implementation of the new, rapidly progressing GIS technology. GC accumulates achievements in the theory, methodology and production practice and, in turn, has a marked influence on the concepts in cartography, the methodological apparatus, and technologies.

Fundamental, real-time cartographic support in resolving problems related to environmental protection and rational nature management on the basis of cartographic-geoinformation research has become a major direction of the Institute's activities in the field of geographic cartography. The geoinformation concept of system ecological cartography has been developed. The major cartographic and geoinformation research areas pursued by Institute involve developing relevant methodologies and mastering current and creating new GIS technologies for the cartographic support in solving problems of regional development.

Such a system has been created at the Institute, permitting a switch-over from conventional map-making to new technologies, digital and computer maps, and atlases. A wealth and the great diversity of means and capabilities provided by information technologies means in addressing general geographical and cartographic issues is taken advantage of through the use of a large number of thematic and integrated as data bases, as well as geoinformation systems of a different intelligent level which are characterized by the presence of interaction of information-simulation models with thematic expert systems, and with visualization and graphical representation capabilities that were developed for representation of geographical phenomena.

The main areas of activity are as follows: 1) elaborating the theory and methods of system analysis of geoimages and of geoinformation map-making for a balanced territorial development; 2) mapping-based assessment of the natural environment state, monitoring and risk of its modification, as well as of the impact of adverse environmental factors on the life of population, and recommendations for prevention and avoidance of their consequences; 3) cartographic representation of resource prerequisites and ecological limitations, and normatives in order to ensure sustainable regional development to identify organizational management mechanisms in nature management rationalization in the region; and 4) addressing problems of normative legal assessment of computer-generated maps and atlases for their efficient practical implementation, including in international cooperative projects.

Within the framework of the above areas, it is envisaged to address interrelated problems through their step-by-step implementation. The following top-priority tasks were defined: 1) the creation of the geoinformation model for a territory containing initial digital and (different-scale) cartographic material concerning thematic and integrated problems of the region, with the development of new territorial mathematical models and adjustment the existing models to existing systems; and 2) summarizing particular experience of fitting the primary model into the technology of developing a documented, reliable and regularly updated problem-oriented (for sustainable development problems) groinformation model of the territory. In addition, it was possible: 1) to generate computer maps in the real-time mode in various practical applications; and 2) to generate the data exchange and transmission communication system with a broad range of users.

A scientific knowledge gained to date from studying and mapping the Lake Baikal region [2], experience in producing large fundamental atlases, the methodology developed for system ecological cartography, for cartographic support of landscape planning and ecological-economic zoning of the territory, and the extensive data base generated in the form of different thematic layers of spatial information make it now possible to summarize and represent all available materials in the form of electronic maps, and to realize the concept of producing a multimedia atlas in the form of a special-purpose atlas information system (AIS).

The region's natural-ecological and socio-economic potential, and the possibilities for current and future prospective use and protection of natural resources in the AIS under development are represented in such a manner that they can be used in developing economic, ecological and social projects and measures. This is favored by the following capabilities of the electronic cartographic product: 1) systemic, balanced and comprehensive display of the evolution-historic, political-legal and national-religious aspects, the natural-resource, economic, demographic and socio-cultural prerequisites, and the ecological and innovative factors of regional development; 2)the great diversity and the multi-scale character of cartographic material which manifest themselves in a combination of a regional, oblast (republican), district and municipal level of data display, and in the display of information about problem nodes and areas; 3) suitability for management (availability of maps for assessment, norm-setting, optimization, planning, monitoring, and control); 4) the capability of information-retrieval functions, and convenient forms of dialogue with the user; 5) the capability of an integrated processing easy adapting of the electronic atlas to a change in the forms of input and output documents; and 7) availability of various data models that are specialized in data processing and utilization tasks which are closely interrelated in the atlas supporting information system.

The AIS "Sustainable Development of the Baikal Region" a sophisticated organizational structure intended to represent dynamical (across time and space) elements, phenomena and conditions of environment, and natural and anthropogenic factors of formation of territorial development systems. The AIS will become the necessary element in the system of real-time information support for addressing problems of sustainable regional development. It combines elements of a fundamental geographical product, reflecting the interrelationship between social production and natural environment in all its complexity, and the interaction of internal and external factors, as well as the dynamical nature of the processes of nature-society interaction. The last task is solved in the AIS by the capability of real-time generation of derivative special maps for the dynamics of phenomena, and various materials displaying the political, economic, resource, social and ecological situation in their evolution.

The AIS "Sustainable Development of the Baikal Region" will be used in a set of developments that determine the strategic directions of nature management and optimization in regard to the development of the Baikal region. It will provide systems of a different management level with a unified information base and will become the necessary element and the normative-legal basis for the definition and pursuance of the regional policy of sustainable development.

The Institute maintains international relations in the field of cartography, primarily with the International Cartographic Association (ICA). Over a period of 4 years (1968–1972), V.B. Sochava chaired the Commission on Thematic Maps of this Association. This promoted the establishment of scientific ties with foreign colleagues, and some contacts on special issues. More specifically, V.B. Sochava developed the program and general structure of national monographs on the present status and prospects of thematic cartography. His proposals were published in the Russian, English and French languages. The Institute's scientists are members of the National Committee of Cartographers of the Russian Federation (NCC) and are involved in Technical Commissions of the International Cartographic Association.

Results of cartographic research on Siberia have been and are discussed at various scientific forums, including specialized ones. For instance, the V.B.Sochava Institute of Geography SB RAS alone has hosted seven scientific cartographic conferences (Table 3). Discussions of current issues at them made an impressive contribution to the advancement of geographic cartography.

Siberian scientists presented their papers to various international scientific forums: to geographical, geological and botanical congresses, ICA conferences, to ecological and economic forums, etc. An important event was the international Conference "Intercarto-2: GIS for the study and mapping of the environment" organized at Irkutsk by the Institute of Geography SB RAS together with Moscow university in 1996, with more than 100 participating scientists and specialists from Russia, France, the USA, Germany, Malaysia, and other countries.

Table 3

Scientific conferences on geographic cartography hosted by the Institute of Geography SB RAS

No.	Conference status and title	Conveners	Year
1	Third All-Union Conference on Thematic Cartography	Institute of Geography of Siberia and the Far East SD AS USSR, Geographical Society of the USSR, SD AS USSR Commission on map-making for environment, economy and population	1968
2	Regional Conference "Cartographic Support of Planning of Territorial-Production Complexes"	Institute of Geography of Siberia and the Far East SD AS USSR SD AS USSR Commission on map-making for environment, economy and population	1977
3	Regional Conference "Cartographic Support of Regional National Economic Programs"	Institute of Geography of Siberia and the Far East SD AS USSR Geographical Society of the USSR (Bureau for Siberian and Far-Eastern Organizations), SD AS USSR Commission on map-making for environment, economy and population	1983
4	Conference with all-Russia and International attendance "Cartographic Methods of Investigating Ecological problems, and Nature Management Optimization in Siberia"	Institute of Geography SD RAS, East-Siberian Division of the Russian Geographical Society, SB RAS Commission on Cartography	1988
5	Conference with all-Russia and international attendance "Resource-Ecological Mapping Based on Information technologies"	Institute of Geography SB RAS, East-Siberian Division of the Russian Geographical Society, SB RAS Commission on Cartography	1993
6	International Conference "InterCarto-2: GIS for the Study and Mapping of Environment"	Institute of Geography SB RAS, M.V. Lomonosov Moscow State University, East-Siberian Division of the Russian Geographical Society	1996

7	VII Conference with all-Russian and international attendance "Geoinformation Mapping for Balanced Territorial Development"	Institute of Geography SB RAS, East-Siberian Division of the Russian Geographical Society	2002
8	VIII Conference with all-Russian and international attendance "Cartographic and Geoinformation Support of Regional Development Management"	Institute of Geography SB RAS, East-Siberian Division of the Russian Geographical Society	Nov. 2006

The modern concept of systemic geographical mapping of Siberia's regions is based on a subject-oriented study and mapping of the key factors that have influence on the achievement of the chief goal of territorial development, the goal to improve the standards and quality of human life. A fundamental cartographic endowment of many regions of Siberia, the Baikal region in particular, provides a means to create fundamental cartographic application products in real time in order to identify in the region territorial zones of growth and development, conservation, and sanitation; to produce the region's cartographic image and carry out investment marketing (to prepare the territory for investments); to create in real time composite applied maps for municipal formations, companies and enterprises; and to create efficient, illustrative cartographic products of educational and training purposes.

The research results of Siberian scientists in the field of geographical cartography, and a further solution of problems, issues and tasks of mapping the vast Siberian and adjacent regions would enhance the role of geographical cartography in addressing current problems of sustainable development of regions.

References

1. Postnikov A.V. The Advancement of Large-Scale Cartography in Russia. Moscow: Nauka, 1989, 229 p.
2. Vorobyev V.V., Batuyev A.R. The contribution of Siberian geographers to developing thematic cartography. Cartography at the Turn of the Millennium: papers of the 1st All-Russia Scientific Conference on Cartography (Moscow, 7-10 October 1997). Moscow, 1997, pp.54-65.
3. Vorobyev V.V., Batuyev A.R., Belov A.V. et al. Ecological Mapping of Siberia. Novosibirsk: Nauka. RAS Siberian Publishing Company, 1996, 279 p.
4. Berlyant A.M. Geoinformation Map-Making. Moscow, 1997, 64 p.

CHAPTER

18

Geosystems Analysis of Siberia's Landscapes using Remote Sensing

A.K. CHERKASHIN, V.A. KUZMIN AND V.M. PLYUSNIN

The specific feature of the geosystems approach in geography involves inter-sectoral, multi-component and dynamic interpretation of geographic data based on studying the typological differentiation of a territory's landscape to unravel the connections characteristic of the various levels of naturally or anthropogenically modified geosystems. This conception of geosystems analyses imposes special requirements on the use of satellite data as the source of objective ideas of the invariant space structure of the areas contoured within landscapes. All this becomes a methodological basis for geoindication and landscape indication, thematic interpretation of satellite images, and the generation of branching technologies for landscape geographical support of territorial development.

Current satellite data have all necessary features for solving fundamental tasks in studying the geosystems' space differentiation at a local or regional levels. Especially valuable are the digital hi-resolution different-range images: MSU-E (Resurs-O, Russia), ETM+ (Landsat, USA), HRV (SPOT, France), LISS-3 (IRS, India), MK-4, KFA-1000 (Resurs-F, Russia), TK-350, and KVR-1000 (Kometa, Russia). They provide data that were obtained previously through field expeditions or station-based investigations only. Detailed studies of the territorial contrasts, delineation of the geosystems' natural boundaries, identification of regular-type connections between the components and the spatial functional conjugation of natural complexes provide a further insight into the structure and dynamics of geosystems.

The V.B.Sochava Institute of geography SB RAS has been and is engaged in the development of models and methods for computer-aided image interpretation. GIS technologies are developed for processing multizone data of Earth remote-sensing and topical mapping for solving various geographic tasks (Fig. 1). Research results and applied products are made use of when conducting the ecological-geographic examination of economic projects, the landscape planning of a territory, land assessment, the functional zoning of specially protected natural territories, the estimation of anthropogenically disturbed natural complexes, of pollutant emissions into the atmosphere by industrial centers.

The software package has been developed, the organization was based on fibering procedures for data, programs and blocks of a process conveyor [2]. This package permits a remotely sensed data processing within any electromagnetic spectral range to complement GIS-projects. It consists of a number of collaterally subordinate modules and enables 12 operation groups to be distinguished: image studies, object and event classification in the visual and automated modes, image synthesis in conventional colors for several frequency ranges, etc. It is used largely to process raster data. As this complex constitutes the software core for computer-aided technology of geoimage interpretation, it can be used to advantage in such applications as processing, and semantic and numerical analysis of remotely sensed data, the synthesis of thematic interpretations of digital images. It can be nicely adapted to a variety of multi-purpose expert examination issues related to regional projects and various remotely sensed materials. Research methods that were developed as part of this study provided an array of practical and scientific results from solving natural history-related problems in many Siberian regions.

The core of most geographic and ecological tasks is formed by territorial (spatial) zoning. As an element for justification, assessment and decision-making, it is included in many normative and legislative acts and generally relies on utilizing landscape patterns, well reflected by remote-sensing data. They permit landscape structure analyses on several hierarchical levels to revealing primary structural patterns and elaborate, in sufficient detail, small but important (in terms of landscape), elements of geosystems.

A number of characteristic properties of landscape structure were identified in southern East Siberia. Most important of them are the combined taiga, tundra and steppe types of natural environments; the complicated character of mutual penetration of different natural systems; a considerable landscape-forming influence of landforms; variations in landscape structure of the macro-slopes determined by the location of the ridges, and by the prevailing eastward transport of moisture; a mosaic natural territorial structure, determining the complexity of landscape, the contrast and the specific character of local variants of the natural background [3].

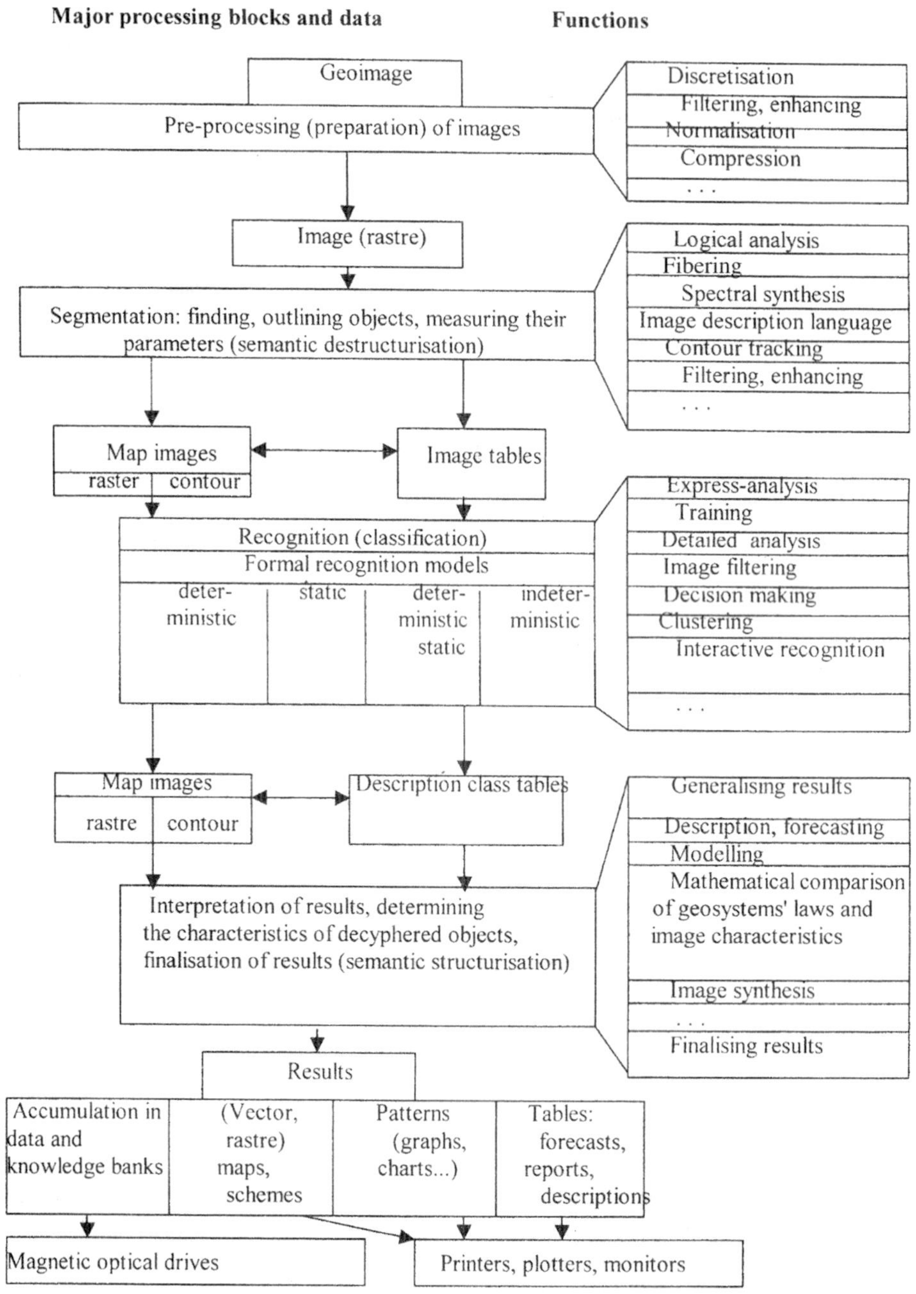

Figure 1: Structure of computer-aided geoimage interpretation [1]

Methodologically, the analyses of landscape structures were performed at various hierarchical levels: planetary, regional and topological. Accordingly, different-scale satellite images were employed: 1:2,500,000 scanner images from MSU-S and MSU-SK, scanner images from the Landsat system, 1:1,000,000 and 1:500,000 (black-and-white multi-zonal and spectral-zonal) images, as well as 1:200,000 and 1:100,000, mostly multi-zonal, images. Landscape analyses using the images involved, first of all, identification of the information content, the landscape structure composition, structural kernels, and critical components and, secondly, the discussion of geometrical components: contour shapes, sizes, contour orientation, mutual location, uniformity, and the degree of contrast.

Fig. 2 presents a structural schematic of the Baikalsky range, a sloping horst, part of the Baikalsky rift zone and its western framing. Here, one can clearly see five first-order structural units extending submeridionally: A – dark grained structure of dark coniferous taiga in the western macroslope of the range; B – mosaic high-altitude alpine-type structure of the range consisting of rocky watersheds, corries, and upper trough valleys, glaciated in the late Pleistocene; V – spotted curvy structure of stone-stream and brush-clad bald peaks that were in periglacial conditions during late-Pleistocene glaciation, in the eastern medium-altitude hills of the range; G – grained structure of light coniferous taiga of the eastern foothills of the range, complicated by light-colored strips and wedges of "steppe-like" forest-free slopes; D – dark flat structure of Lake Baikal.

Second-order structures include forested trough valleys (1), penetrating into the high-altitude mountains of the range and reaching the foothills as curvy moraine complexes (2) of a compact oval and radial shapes. Clearly light lines in the western macroslope's foothills are the "mari"* (3), treeless meadow areas on permafrost ground along the valley floors. The linearly striped, sometimes, fan-like texture on steep slopes is formed by erosion rills (4). A dark grained structure (5) is created by abies-cedar small-brush, small-grass, green-moss forests. Light patches with clear boundaries correspond to burned-over areas, some of which are already getting grown over by birch and aspen-trees and are somewhat darker in color (6). The cellular structure is made up of corries and upper trough valleys (7). Along the corrie bottoms, dark patches of alpine lakes are frequently seen (8). Along the river valleys, moraine-dammed lakes can be found (9). This image clearly shows compositional differences in the mountain taiga between the western and eastern macroslopes of the range. While the western part is dominated by dark, almost maroon, shades of dark coniferous taiga, the eastern part is dominated by the lighter-colored larch and pine-larch taiga, often with Daurian rhododendron in the underbrush (10). All along Lake Baikal's shore there are light-colored strips and wedges of "steppe-like" treeless slopes (11). They typically have a southern or south-eastern exposition. Stone-stream and shrub-grown bald peaks (12) on flat or slightly-sloping mountain-top surfaces form a grained pinkish texture divided by greenish elongated patches of mountain taiga forests of reduced development. Some places along the Lake

Baikal shore have a clearly visible narrow light-colored strip of sand-pebble sediments, while on some capes, on alluvial cones, dammed "sor" lakes have formed (13).

Landscape differentiation on the ridge is more pronounced in images taken in the spectral range of 0.5–0.6 um and 0.6–0.7 um, where the color intensity range is maximal, while the outlines of landscape contours reflect most distinctly the differences between forest (dark) and bald-peak (light-colored) areas.

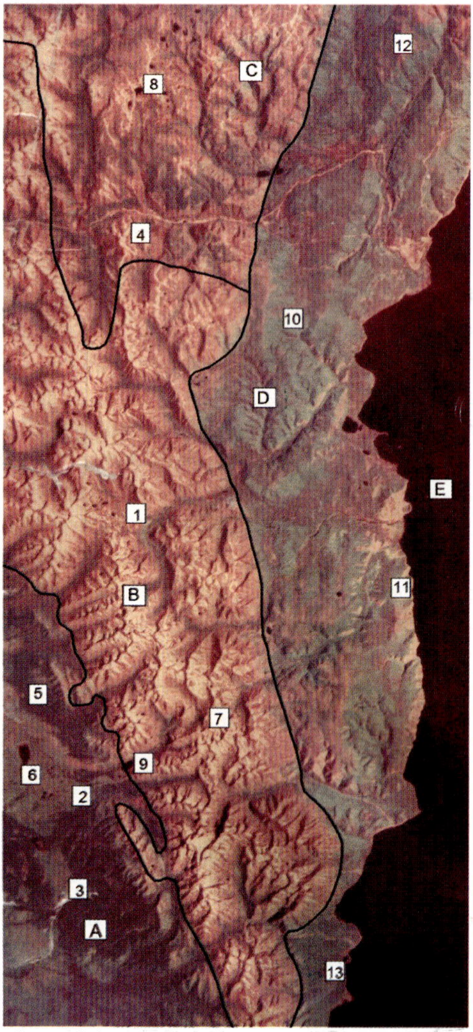

Figure 2: Structure plan of the Baikalsky range (from a 1:500,000 satellite image)

The Prebaikalia's depression is dominated by foot-hill dark coniferous geosystems on glaciation-related sediments or on precipitation carbonate earth-materials, respectively, with acidic coarse-humus soils, often with a differentiated profile, or with turf-carbonate, turf-podzolic residual-carbonate soils. No apparent links have been discovered between them and the images and vegetation. This is partly explained by fire-disturbed forests, which complicates the reconstruction of their native vegetation from images. Morains have the shape of hill-ridges. Earth materials of heavy granulometric composition or with near-surface permafrost have rills with stunted-tree spinneys or grooves of sparsely scattered larches (lighter-colored in images) which consist permafrost swamp soils, and peat-mould permafrost soils, while forested hills are composed by podsols. Lightweight earth matrials with good drainage have a less contrasting soil cover. Typically they appear as combinations of podzols and brown forest coarse-humus soils.

Widespread along the Baikalsky range are bald-peak alpine-type geosystems, bald-peak tundra and l shrub-clad piedmont geosystems, as well as dark coniferous sparse-forest piedmont geosystems with tundra, podzolic soils, and podburs. The eastern hoothills of the range are characterized by a predominance of mountain-taiga larch geosystems of optimal or limited development on flat surfaces and along low-grade slopes with turf podzolic, turf forest soils, and podburs.

The study of the geosystems dynamics using satellite images included four methodological techniques [4]. One involved comparing images taken at different times. This method is widely employed by researchers, as it is well developed; it is used mainly in analyzing the dynamics of particular components of geosystems or natural phenomena. Comparison of images taken at different times is done visually or instrumentally. Information on changes in objects could be obtained through a comparison not only of different-time images, but also of images with maps, images with natural objects, images with drawings, and images with the geographical description of an object. The second methodological approach involves in analyzing aerospace images to be used in regional landscape forecasting and in reconstructing the old landscape structures. Images are assessed not only as spatial models but also as spatial-temporal landscape models. Once traces of landscapes that existed at some older time (tens, hundreds or even thousands of years ago) are detected, their boundaries can be outlined, while indirect interpretation evidence could also help to identify relevant geosystems. The third technique implies identifying current dynamic status of and development trends in geosystems. Their status is understood as the relationship of their structural and functioning parameters under the impact of external and internal factors over a certain time period: 24 hours, a week, 10 days, a month, a season, a year, and several years. This method is only poorly elaborated and demands further scientific research. Geosystem state variables can also be diagnosed from images in the analysis of their dependencies on local (both natural and anthropogenic) environments. In the process, links are revealed, which determine the trend and the degree of internal completeness. This implies a fourth methodological technique, i.e. that of analyzing the dynamic behavior of geosystems in relation to computer-aided processing of images. Computer-aided interpretation of such elements as the number of contours per unit area, ragged borders, dimensions and shapes of contours, occurrence of contrast images, the sharpness of contour borders, and a number of other parameters, enable quantitative indices to be obtained and interpreted as dynamic properties of geosystems.

Remote-sensing of disturbances in Siberian landscapes has shown that anthropogenic factors emerged two or three hundred years ago, yet their load on geosystems is comparable to natural stress. Satellite images taken in the winter time show trace dark spots of dust coming from industrial enterprises around many cities. The halos of polluted snow surrounding urban centers are several times as large as their own areas. Forest clearings are quite well discernible against the background of forests by their intrinsic geometrical shape and high albedo. The boundaries of old clearings that are starting to be grown over with vegetation, are less clearly visible, but they are also discernible thanks to different graininess of their structure. The boundaries of burned-over areas are tongue-shaped, often coinciding with natural borders: rivers, lakes, and gullies. Occasionally, mining operations transform landscapes to technogenic badlands.

Measures have to be taken in order to preserve natural landscapes, and their ecological stability. For this purpose, specially protected natural territories have been established: reserves, national parks, refuges, and natural monuments. In addition to creating a network of specially protected territories, other measures are necessary to control the geoecological equilibrium. They include the creation of anti-erosion and anti-mudflow facilities, restoration of previous forest geosystems in areas of tree felling and in burned-over areas, and recultivation of areas occupied by spoil heaps, with due regard for the landscape patterns in mountains in the case of a particular project.

Models of ecologically safe sustainable economic development were developed for the Lake Baikal region. Given the adoption in 1999 of the Act on Lake Baikal, the leading role is to be played by nature conservancy measures, with the recreation sector set aside the top-priority economic sector. Criteria of ecological stability and diversity of geosystems were defined, as well as their ecological values. This implies their environment-forming and stabilizing function. All measures aimed at building the recreation infrastructure must comply with the specific features of particular localities, according to the "do not harm, but improve" principle. The present ecological status of a territory should be regarded, primarily, in terms of disturbed natural landscapes, negative effects of economic activities, as well as the resistance of natural complexes to anthropogenic stresses. This is manifest in the self-purification ability of geosystems, depending on the intensity of geochemical, geophysical, hydrological and climatic processes.

Assessments of anthropogenic impacts on the natural environment should be based on identifying the types and degrees of such impacts, as well as on the geosystems' resilience to different types of loads and self-purification abilities. Methodologically, the resolution of this problem breaks down into 3 parts. In the first place, aerospace images help to reveal the types of disturbances in geosystems. The available literature and reference data supplement the "impacts" module with those not visible directly in images (pollution of air and water environments with toxic substances, and impacts from local points) thus helping to determine the time and periodicity of impacts and calculate the disturbance level. The second part involves estimating the resistance of geosystems to anthropogenic loads. Landscape resistance implies the ability to retain their structure across space and time under changing environment conditions. Resistance is related to the dynamic properties of geosystems. Indigenous facieses are more resistant (resilient) to loads than serial ones. Therefore, landscape mapping of a territory must identify, first of all, indigenous as well as other facieses, including the various derivatives (anthropogenically altered ones) [5]. They must form the framework for identifying the geosystems' resilience to loads. The third part of solving the task of assessing anthropogenic impacts implies determining the wind regime, existing permafrost, temperature inversions that make the vertical transfer and dispersal of pollution difficult, the filtering properties of earth materials, hydrological regimes of rivers, the amount and rate of atmospheric precipitation, the specific amounts of surface discharge, etc., encouraging the dispersal, dilution, removal, assimilation, recycling of external technogenic pollution [6].

A methodology of polysystem analysis and synthesis of geographic objects has been developed which basically involves stratifying complex objects into a set of interpretations regarding a non-overlapping system of objects. For each of them, an axiomatic theory is created as well as a set of mathematical research models [7]. In processing satellite data in order to tackle particular geographic problems, an appropriate model is selected for processing remote-sensing data. For instance, complex-systems quantitative-theory models are used in developing methods for assessing natural resources based on multi-zone satellite images. Models of macro-dynamic processes are used to investigate the geoimage properties and solve the geoindication problems. The poly-system methodology makes it possible to regularize different geographic patterns. Each pattern is assigned its own place in the system of theoretical knowledge, as well as new models and methods are provided for processing field and remote-sensing data. The methodology of landscape synthesis of geographic knowledge is built upon the ideas of geoinformation support. They have their origins in the study of the general principles by which to generate and utilize a geographic-science knowledge and switch over to a representation of its methods and the results from using them and, subsequently, to address specific tasks posed by nature management [8]. This resulted in the definition of the principles and methods of landscape classification and nature regionalization, e.g., a variety of combinatory classifications of geosystems in West Siberia were discussed. Particular examples of geographic research in various regions of Siberia help to reveal the contents of the general principles of predicting ecological conflicts and methods to solve them, specifically, the problem of land aridization. Major patterns of natural environment differentiation were presented, as well as criteria for ranking landscapes as specially protected territories.

Long-term research implies developing objective automated procedures for compiling landscape typological maps based on multi-zonal satellite data, and building the geographical information support based on Earth remote-sensing data, knowledge bases, and GIS technologies. Combined satellite data, on-land station-based and expedition research results, combined with the researchers' experience, hold promise for switch-over to novel theoretical concepts of the spatial organization of natural environment and give an opportunity to create methods of image contents interpretation.

References

1. Kitov A.D. Geoimage processing systems as used in solving various geographic problems. Geografiya i prirod. resursy. 1998, No. 4, pp. 103-112.
2. Kitov A.D. Computer Analysis and Synthesis of Geoimages. Novosibirsk: SB RAS Publishers, 2000, 220 p.
3. Mikheev V.S. Landscape Geographic Support for Siberia's Complex Problems. Novosibirsk: Nauka, 1987, 208 p.
4. Plyusnin V.M. Aerial and satellite photographs in exploring the geosystems' dynamics. Geografiya i prirod. resursy, 1992, No. 4, pp. 140-146.
5. Sochava V.B. Introduction to the Doctrine of Geosystems. Novosibirsk: Nauka, 1978, 320 p.
6. Bolshakov A.G. Ecological bases for planning the Lake Baikal region. Nature Management in the Lake Baikal Watershed Basin. Irkutsk, 1988., pp. 139-151.
7. Cherkashin A.K. Polysystem Analysis and Synthesis. Implementation in Geography. Novosibirsk: Nauka, 1997, 501 p.
8. Mikheev V.S. Landscape Synthesis of Geographic Knowledge. Novosibirsk: Nauka, 2001, 215 p.

CHAPTER

19

Geographical Science and Ecological Education as Conditions of Development of A Civil Society

I.V. KONEVA

The subject matter of geographical science is becoming its interaction with ecological education. The reasons behind this interaction and its essence are determined by logic of the advancement of science and education.

The processes occurring in a separate branch of science cannot be understood without regard for processes occurring in science in general. The specific function of science as a form of social consciousness is known to involve working out the theoretical picture of the world and providing an adequate image of reality which, in turn, is only a means, condition, prerequisite and foundation for pursuing man's transforming activity. The socio-cultural role of science resides in the fact that a rationalization of man's activity is accomplished. Science works out and offers to the practice theoretically grounded ideal plans and programs of activity. Their practical implementation always involves a need to take into account the particular circumstances, and invoking personal activity and initiative of the actor. The role of education involves reproducing the ideal plans and programs of activity that are offered by science, adapting them to the realia, teaching people a habit of transforming ideal plans into concrete plans and implementing them. This process incorporates not only higher but also general education. It is precisely this education which is able to promote self-government, based on assimilating algorithms and programs generated by science.

The expansion of the functions of science is associated with the evolution of its ideals and criteria of progress. An ideal of science implies constant targeting at revealing an objective, regular interrelationship and mutual conditionality of phenomena of the material and spiritual world (formation of the picture of the world). This ideal, in turn, is not a stiff one. The overall tendency of its transformation is toward a constant complication of the picture of the world. Among the tools used to generate a unified scientific picture of the world are: the integrated character of research, system organization of scientific knowledge, and ecologization of science.

Humanism came to be a criterion for progress in the 20th century – the emergence of man in the evolving picture of the world. There was a need to reveal a linkage of naturally occurring and social phenomena, and to look for how the advancements of natural science and social science can be joined in the picture of the world. That resulted in a change of the notion of the relation of the attribute of fundamental science uniquely to sciences treating on Nature. The right to play a role in the creation of a unified picture of the world is being acquired by world-out-look related sciences, systems of sciences, and by interdisciplinary fields of science. The problem of man-habitat interaction (Weltanschauung issues) is becoming a central problem of science and the measure of its fundamental character.

One further criterion for progress of science in the 20th century involves the emergence of ideas about the ultimate ideal of science – expedient activity, and its outcome: the economic, ecological, social and cultural effect of science. The notion of activity in modern cognition s coming into particular importance. In this notion, scientific cognition merges with the implementation of science, with the transformation of the world. Along with problem-oriented research, in science there are emerging problem-oriented research that is becoming dominant. In this connection there arises a need to broaden the gamut of interdisciplinary research, create the methodology for such research, and to alter the views regarding the end product of science. As such, we have programs, algorithms, and technologies. Among social technologies are also the technologies in the sphere of education.

The system of geographical sciences that combines into a single block natural and social sciences, can realistically meet today the requirements of society for the creation of the requisite scientific picture of the world and make its decisive contribution to the formation of a scientific world-outlook. It has a key role in the reforming of education in accordance with the challenges of the present time. The best chances for integration of a knowledge of Nature and society are known to be afforded by regional investigations. The advancement of geography since the mid-1950s of the past century has been determined by the ecological vector, the methodological essence of which involves integrating a knowledge.

The methodology of geographical synthesis which has been developed by the Institute of Geography SB RAS over the course of nearly half a century of its existence, can render this activity successful. The ecological orientation in the activity pursued by the USSR AS Institute of Geography of Siberia and the Far East (later – Institute of Geography SB RAS), contributed to the development of the geosystem concept (landscape ecology). Likewise, it determined the shift of geography to human ecology which, according to V.B. Sochava [1], is the highest analytical stage of essentially unified geographical research. Today, in reply to the challenges of the time, on the basis of geography as human ecology, the Institute is evolving the concept of ecological education oriented to the ideas of sustainable development.

The thrust vector along the line of integration of geographical knowledge was defined as early as the period of Institute's establishment by V.B. Sochava's ideas: about the subject of geography as a geographical environment, about the integrated character of future research, and about the development of Siberia as an ideal of the ultimate aim pursued by geography. On the whole, these three components can be regarded as a unity of the goal, the means and the result.

In Soviet times, integrated activities of Siberian geographers were in line with the scope of state activities and were associated with development of oil and gas fields in West Siberia, the development of the Angara-Yenisei region, and with the construction of the Baikal-Amur railway. They were submitted, as scientific information for management, to such agencies as Gosplan, Gosstroi, the State Committee on Science and technology, and others, and were used in these institutions and organizations in planned and projected developments. The form and style of such geographical work were determined by the "specialist-specialist" interaction.

In post-Soviet times, the Institute became engaged on research of regional and local nature. That research had not so much a unique as standard character. The addressee (especially at a local level) was then represented not as much specialist geographers as the local community. In such circumstances it was necessary to complete the research by working out plans or projects which represented the product approaching what could be called an ideal of the ultimate goal. Addressing standard problems and the general consumer necessitated developing algorithms for planning and designing activity. It was necessary to objectify the designing process: to divide the designing task into parts, specify the interrelations between these parts, and create a feeling of thinking over the design "showing one's hand" in order that consumers could participate in the process of designing and discussion. Such a goal was achieved through the numerous plans for ecologically oriented land use executed for different territories of the Baikal region. However, even such "adaptation" to the perception by the general consumer of products of a research team was insufficient. The problem may be interpreted as the society's non-readiness to participate in discussions, as the keeping-aloof from the solution of societal problems. The solution of this problem resides in the sphere of education through the development of public consciousness, and the scientific world outlook. The scope of this problem goes far beyond the interaction of geographical science and society in a separate region. It is reflected in many well-known international documents.

In particular, one of the reports prepared at the request of the Rome Club by a group of authors including representatives of the USA, Marocco and Romania, was devoted to problems of public education and was entitled: "There are no limits to education". In analyzing the education problems, the authors of the report concluded that the existing system of public education, if a global scale is kept in mind, has become anachronic, inapt to aid in resolving the challenging problems facing mankind. They recommended that the system of public education should be radically reformed and oriented at urgent current problems facing mankind, at an understanding of the general global character of these problems, and at the improvement of the teaching process (by introducing new, more progressive, methods).

An important significance was attached to education in the Concept of Sustainable Development that was adopted as the agenda for the 21st century at the UN Conference on Environment and Development held in Rio de Janeiro in 1992. Contemporaneous education had to transform the Concept of Sustainable Development into a system of spiritual and professional directives of mankind. A central role in that program would be played by the sphere of education which in the English speaking countries is termed environmental and in this country – ecological. This target designation is reflected in recommendations of various international congresses and conferences on education.

There is also a growing awareness in the Russian general public of the fact that the movement into a post-industrial society dictates the need to revise the content of education and its technology, and to refine the optimal structure of knowledge both for the young generation and for adult population. A goal-oriented alteration to the content of education, as opined by specialists in the field. The reason is that trainees do not receive the necessary (in modern society) information at every stage of education, starting from school of general education. Knowledge must be directed so as to meet the requirements of a civil society and reflect those problems which agitate society. It is anticipated that education (especially at school level) must be directed not so much to the problems facing society as to their awareness. Such education would receive, to some extent, a problem orientation. Analyzing the school educational standard that is formed, specialists do not find in it even such fundamental notions upon which we attempt to build a civil society. The old system of education was adapted to meet the requirements of a militarized, autarchic country. It was good in its own way. But a new system of education will have to be adjusted to other problems. Whilst objective processes are of paramount importance, modern societies often face the fact that the objective factor starts to play an increasingly larger role. A significant and, often, determining factor is becoming the influence of social consciousness which, to an ever greater extent, is the indicator and criterion of the changes that are taking place. Changes in the world are accomplished not on barricades, not in the course of wars and other forms of violence but under the influence of what happens in human hearts and souls or, to put it otherwise, in social consciousness. The failures of Russian reforms in the 20th century are attributed by specialist sociologists to the indifference of most society. The problem of our school resides in the fact that it does not teach human beings to adapt themselves to the hard modern reality and form a fundamentally new attitude to life. One further problem is that today the heads of general-education school leavers create, as specialists figuratively put it, the situation in the form of isolated shelves of a rack making accord usage of different kinds of knowledge difficult. The appearance of ever new isolated shelves in the heads of grown-ups may well be regarded as a hazardous disease, to combat which is among the problems of the state academic policy.

The chief cause for the crisis of the system of education, in the view of specialists, is associated with lack of student's activity. The challenge to education on the part of society involves not providing the trainee with a still greater amount of information but teaching her/him to play and win in his/her life, taking advantage of information as an additional mechanism in order to be the gainer. One way to resolve this problem is by moving from the subject-lesson system of education to designing activity, immersion of the student into the problem. Such activity would permit arranging a knowledge obtained through subject teaching into a system oriented at the ultimate, socially useful goal.

Man is affected most seriously by those problems which immediately surround him. Such are the problems of a regional and local character. To make sense of them and get acquainted with the algorithm for solving these problems is the condition making for cultivation of an intelligent attitude to life, the making of a civic stand of man, the formation of a civic society, the development of self-government, and the advance in line with the strategy of sustainable development. Ecological education that builds upon the problem-oriented basis, refers to the interdisciplinary realm, human ecology (social ecology). At the present time it goes also under the name of ecological education for sustainable development. Most realistically, its advancement is possible on the basis of the regional component of school education. In a more thorough scope, such education for sustainable development is possible and especially needed in rural schools where all components of sustainable development: bioecological, social, and economic – are interrelated on a limited territory in the immediate vicinity from students, while the indifference of rural residents in regard to the present reforms is particularly conspicuous and presents a serious problem. Ideally, every rural school must have its own project of development, identifying the problems of the territory (bioecological, economic, and social), taking into account the resources and limitations, planning and implementing activities on the development of the territory, setting up a monitoring of the advance to the goal. Such a project should be prepared, implemented and gradually developed by the team of teachers and students, with the involvement and collaboration of the administration, sectoral structures, and all residents of the territory. Rural school can become a consolidating element in uniting the local community, in its organization, and in the development of elements of a civic society. The geography teacher with a training background sufficient for carrying out designing activities must be able to develop – using locally accessible material (forest taxation data, land use maps, and topographic maps) at least two maps: the map for actual utilization of the territory, and the geobotanical map (schematic). These, in turn, would serve as a basis for forming the regime of nature management and working out the program of actions and activities on the development of the territory. The history teacher (with the help of students) would reconstruct the history of the district; the chemistry teacher would evaluate the quality of agricultural lands and pollution of the territory; the social science teacher would analyze the socio-economic situation; the literature teacher would participate in the drawing up of the project and the creation of the document fostering the image of the territory, and the foreign language and informatics teacher would tackle communication issues: access to the Internet, and communication with like-minded persons in international projects.

Institutions of higher education specializing in training teachers are to teach school teachers to manage project activity. That would require some rearrangement in the activities of institutions of higher education. Interdisciplinary project activity must become the subject of their special care. A great deal of assistance in this regard could be rendered by geographical science that arranges its activity on an interdisciplinary basis in the ecological profile that evolves in accordance with criteria of humanism and ideals of sustainable development. Ideal plans to be used, which are worked out by geographical science, could include algorithms of regional, ecologically-oriented planning of land use, supplemented by plans for socio-economic development of the territory. Integration of the Academy's science and of higher and general-education school in the activity on reforming modern education in line with the challenges of the time, complies with the objectives defined by the Declaration about science and use of scientific knowledge adopted at the World Conference on Science held in Budapest on 2 June 1999. That summary document declared the responsibility of science for universal scientific education in a broad sense and on all levels, which is prerequisite for democracy and sustainable development, and for expanding the involvement of population in taking decisions based on the application of a new knowledge. The document recognizes the need to promote the dialogue between scientists and society.

The advancement of ecological education for sustainable development is currently ensured by regional laws about ecological education. Specifically, such a law that orientates ecological education at the goal□ sustainable development of the territory, was adopted in late 2003 in the Irkutsk region.

REFERENCE

1. Sochava V.B. Geography and Ecology (Proc. 5th Congress of the USSR Geographical Society). Leningrad, 1970, 22 p.